# The TEACHING ASSISTANT'S GUIDE

Essential Skills for College Lectures and Labs

**ALSO FROM COLD SPRING HARBOR LABORATORY PRESS**

*At The Bench: A Laboratory Navigator*, Updated Edition

*Lab Math: A Handbook of Measurements, Calculations, and Other Quantitative Skills for Use at the Bench*, Second Edition

*Lab Dynamics: Management and Leadership Skills for Scientists*, Third Edition

*Career Options for Biomedical Scientists*

*Experimental Design for Biologists*, Second Edition

# The TEACHING ASSISTANT'S GUIDE

## Essential Skills for College Lectures and Labs

**ED HIMELBLAU**
*California Polytechnic State University*

COLD SPRING HARBOR LABORATORY PRESS
Cold Spring Harbor, New York • www.cshlpress.org

The Teaching Assistant's Guide: Essential Skills for College Lectures and Labs

All rights reserved.
© 2024 by Cold Spring Harbor Laboratory Press, Cold Spring Harbor, New York

| | |
|---|---|
| Publisher | John Inglis |
| Project Manager | Barbara Acosta |
| Editorial Assistant | Danett Gil |
| Permissions Coordinator | Carol Brown |
| Production Editor | Kathleen Bubbeo |
| Production Manager | Denise Weiss |

**Front cover:** Cover design and illustration by Ed Himelblau.

Library of Congress Cataloging-in-Publication Data

Names: Himelblau, Ed, 1969- author.
Title: The teaching assistant's guide : essential skills for college lectures and labs / Ed Himelblau, California Polytechnic State University.
Description: Cold Spring Harbor, New York. : Cold Spring Harbor Laboratory Press, [2024] | Includes bibliographical references and index. | Summary: "Every year, thousands of graduate students become college teachers. They typically enter the classroom with minimal training. This volume provides practical, accessible advice and guidance for new Teaching Assistants (TAs) with special attention to those who will teach science (labs and discussion sections)"-- Provided by publisher.
Identifiers: LCCN 2024000636 (print) | LCCN 2024000637 (ebook) | ISBN 9781621825173 (paperback) | ISBN 9781621825180 (epub)
Subjects: LCSH: Graduate teaching assistants--Training of--United States--Handbooks, manuals, etc. | Science--Study and teaching (Higher)--United States.
Classification: LCC LB1738 .H56 2024  (print) | LCC LB1738  (ebook) | DDC 378.1/250973--dc23/eng/20240207
LC record available at https://lccn.loc.gov/2024000636
LC ebook record available at https://lccn.loc.gov/2024000637

All World Wide Web addresses are accurate to the best of our knowledge at the time of printing.

Authorization to photocopy items for internal or personal use, or the internal or personal use of specific clients, is granted by Cold Spring Harbor Laboratory Press, provided that the appropriate fee is paid directly to the Copyright Clearance Center (CCC). Write or call CCC at 222 Rosewood Drive, Danvers, MA 01923 (978-750-8400) for information about fees and regulations. Prior to photocopying items for educational classroom use, contact CCC at the above address. Additional information on CCC can be obtained at CCC Online at www.copyright.com.

For a complete catalog of all Cold Spring Harbor Laboratory Press publications, visit our website at www.cshlpress.org.

*Dedicated to Kit, with love*

Visit TheTAsGuide.com for free resources including a curriculum for each chapter and downloadable templates.

TheTAsGuide.com

# Contents

| | |
|---|---|
| Preface | ix |
| Acknowledgments | xi |
| Introduction | 1 |
| 1 Boundaries: Make and Maintain Them | 5 |
| 2 Your First Class | 13 |
| 3 Reflection: Capture Your Experience | 25 |
| 4 Content: Be a Guide for Students | 31 |
| 5 Inclusivity: Help Students Feel Welcome | 43 |
| 6 Participation: Promote Active Learning | 59 |
| 7 Think-Pair-Share: Get Them Thinking and Talking | 75 |
| 8 Grading: Make it Fair and Efficient | 85 |
| Appendix: Guidelines for Teaching | 103 |
| Your Teaching Reflections | 115 |
| Bibliography | 131 |
| Index | 135 |
| About the Author | 139 |

# Preface

### "What if my students don't respect me?"

This question came from a smart, accomplished young woman. She was the first in her family to earn a degree—a Bachelor of Science in Biology—and was entering the master's program to study ecology. She was also about to teach her first class—a lab section that was part of a large introductory biology course—and she was anxious, leading her to question her qualifications and abilities.

I was leading a graduate seminar intended for new teaching assistants (TAs). We had just finished our first session and I thought it went well. I had covered how many office hours to have, how to access your course web page and roster, how to deal with students trying to "crash" the class, and institutional policies about privacy, discrimination, and amorous relationships. I had projected images of my own teaching notes and described my own preparation for teaching. A few brave students had volunteered to share some of their first-day concerns, and I had done my best to address those concerns.

When the student approached me after class with her question about respect, I was surprised. I knew she had a lot to offer her students as a guide and a role model. I knew that she had achieved a high level of academic success and had content knowledge that far exceeded what would be covered in an introductory level course. I knew she had observed many different biology teachers and participated in many lab classes during her years as an undergraduate. Yet her knowledge and experience hadn't translated into confidence in her ability to teach a class. I hoped to respond to her question about respect with something reassuring and helpful, but I was at a loss to answer.

I realize how naïve it was of me not to anticipate the extent to which young graduate students would doubt their ability to succeed as teachers. No one is born knowing how to teach a college-level class. It is not trivial to jump from being a student in a discipline to being a *teacher* in that discipline. And then there's the age factor; many of these new TAs were only a year or two older than the students they would teach, adding a layer of social complexity on top of everything else. It's hardly surprising that imposter feelings would take hold as these young TAs prepared to teach their first college classes.

The students in my seminar had questions and concerns about their new roles as teachers that went beyond the details of course rosters and institutional policies—they needed advice on more pressing concerns. They worried about being unable to answer questions from their students, and how they would recover if they made a mistake. They worried about how to teach a subject for which they felt they had little or no expertise. They wanted guidance on navigating the moment-to-moment interactions that come up during teaching—what the appropriate boundaries were and how to maintain them. They worried about how to balance the time commitments of teaching with their other commitments—research and their own classes. As they gained experience, their concerns evolved. They wished their classes were more interactive. They worried about how to help students who stopped showing up for their class or who were struggling with the material.

I'm grateful to the students in my seminar who raised these important concerns. The material for this book emerged from my attempts to provide help and guidance for my current and future students. The suggestions presented here reflect my own ideas about teaching, the literature on teaching and learning, and insights from observing and interviewing experienced TAs.

There are many excellent books about college teaching—and journals, *and* conferences. But these are mostly intended for faculty. Faculty and TAs are positioned differently relative to the students they teach due to age, experience, and role within the university. TAs need a different sort of book for several reasons: TAs are younger and, therefore, closer in age and experience to their students; TAs are usually not creating a course and are instead delivering a course developed by someone else; TAs simultaneously occupy multiple territories at the university—teacher, researcher, and student—and the boundaries between these territories can blur; TAs teach primarily in small-group settings like laboratories or discussions; and TAs are very likely to be novice teachers.

I've enjoyed the challenge of writing a book of advice for TAs that take these factors into account. Teaching style is personal. I acknowledge that the advice in *The Teaching Assistant's Guide* is influenced by my personal teaching philosophy and might not resonate with everyone. Because there's no right way to teach, I'm sure some readers will disagree with some of my ideas and suggestions. If criticism of the ideas presented here leads to productive discussions about teaching, that's great. Ultimately, I hope readers—especially those about to teach their first college-level class—find something in this book that they can use to help themselves and their students thrive.

# Acknowledgments

Thank you to John Inglis, Barb Acosta, Kathy Bubbeo, Denise Weiss, Danett Gil, and everyone at Cold Spring Harbor Laboratory Press who helped make this book. I appreciate your dedication, your attention to detail, and your positive attitude.

I'm grateful to friends and colleagues who read drafts and provided feedback: Jamie Bunting, Elena Keeling, Annie Meeder, Emily Neal, Harriet Schwartz, and Francis Villablanca. Magdalene Lo was an invaluable editor, helping to get the manuscript ready to go. Finally, I thank the graduate students in the Biological Sciences Program at California Polytechnic State University. The experiences of these students motivated me to write this book and drove many decisions about its content. These young college teachers shared their experiences (several of them are quoted throughout the book), provided feedback, and tested curriculum. I'm very grateful to you all!

# Introduction

THANK YOU FOR TEACHING! Teaching Assistants (TAs) are an important part of the academic experience of most college students. Interacting with TAs in labs and discussion sections can be the most personal academic experience students have in college, especially at campuses with large lecture classes that can feel impersonal.

Good teaching helps students become critical thinkers. Good teaching helps students develop skills they will use in school and beyond. Good teaching promotes equity and access ensuring that all students have chances to learn and succeed. Through your teaching, you will share ideas and information that are important and interesting.

From a career standpoint, the benefit of teaching experience is obvious if you plan to stay in academia where leading lectures and labs is part of the job description. If your career plans are not in academia, you will still use your teaching experience as you train new colleagues or explain your work to a broad audience. For many TAs, teaching is part of the financial equation that supports their education or supplements their income—it helps keep the lights on! The work you do as a TA will benefit you *and* the students you teach.

***Have a positive mindset:*** No one is born knowing how to teach a college class. Teaching is a skill—a set of skills, really—that you learn and hone with practice.

MAGGIE, TEACHING PRODIGY, AGE 3

If you are like most new teachers, there will be some teaching skills that come easily and some that will be harder to acquire. You will try some things in class that work well and some that do not work and that you can improve in future class meetings.

Even if you have never taught before, it is important to remember that you aren't starting from zero. You have been observing teachers throughout your education and that experience, good and bad—but hopefully mostly good—will inform your teaching.

It is normal to feel anxious prior to the start of your first teaching experience. There's nothing I can write here that will completely relieve that anxiety. However, the goal of this book is to give you some of the tools you need *right now* to get your class off to a positive start.

***Who this book is for:*** This book is intended for first-time college Teaching Assistants,[1] many of whom are graduate students. A 2022 survey by the Bureau of Labor Statistics found that more than 135,000 graduate students were employed as TAs in the United States. If we assume that each of these TAs interacts with 20–100 students in a year, we realize that the reach of all these teachers is millions of college students annually. Of course, not all first-time teachers are graduate students. You may be a recently hired part-time faculty member, a postdoctoral researcher, or an advanced undergraduate student. The goal of this book is to help you get off to a good start in your teaching.

---

[1] TAs go by different titles at different institutions such as Teaching Associates, Teaching Fellows, Learning Support Associates (LSAs), Instructional Associates (IAs), Higher-Level Teaching Associates (HLTAs), and others. In this book, "TA" will be used to refer to all these positions.

This book focuses on teaching in a setting like a science lab or a discussion section. These are the settings where TAs are most likely to find themselves. If you are teaching in a different setting, like a large lecture class, many of the ideas and strategies described here will still be useful.

Many new TAs are first-year graduate students facing several new experiences at once. You may be taking graduate-level courses, getting your thesis research started, and adjusting to life in a new place with new people around you. Add the challenge of teaching a college class, and it is enough to overwhelm anyone. It is not surprising that many TAs experience feelings of stress from their simultaneous roles of student, researcher, and teacher (Mazuka 2009).

"WELCOME TO GRADUATE SCHOOL! BY THE WAY, YOU'RE A TEACHER NOW TOO."

If you feel overwhelmed and a bit intimidated by the prospect of teaching, you aren't alone. Hopefully, this book will be a resource to help you ease into teaching by sharing strategies and insights from education research, and the practical experiences of many new TAs. Throughout the book, you will see quotes from TAs who were once exactly where you are now. They were also anxious about stepping in front of the classroom for the first time. They all got through it and went on to become successful teachers. You will too.

This book is short and practical. There is information you need *right now* to help your teaching get off to a strong start. There is information that you might revisit once you have more experience. This book will help you recoup the initial time investment that you will make in your teaching now so that you will not have to spend as much time and energy preparing to teach in the future.

We start with the importance of boundaries in your interactions with students and describe how well-defined boundaries foster a positive, respectful class (Chapter 1). Next are the three most important things to do as you prepare for your first class (Chapter 2), and what to do *after* that first class to make future teaching easier (Chapter 3). Chapter 4 addresses a common concern among new TAs—how to prepare for the content you will be teaching. This chapter also lets you in on a secret: Students might learn a subject *better* when their TA isn't an expert. The next chapters focus on making the classroom more inclusive and welcoming (Chapter 5) and more interactive with strategies that encourage all students to participate (Chapters 6 and 7). Chapter 8 is about writing good assessments (quizzes, exams, etc.) and grading without having to pull all-nighters. Finally, there is an Appendix with information about your rights and responsibilities as a TA.

**HOW HAS BEING A TA BENEFITED YOU PERSONALLY OR PROFESSIONALLY?**

> "I'VE GAINED A TON OF CONFIDENCE—LEARNING HOW TO SPEAK IN A COHERENT WAY TO GET PEOPLE EXCITED ABOUT A SUBJECT I LOVE. TEACHING HAS BECOME MY SAFE PLACE OVER THE COURSE OF GRADUATE SCHOOL. I KNOW THAT I CAN GO [IN THE CLASSROOM] AND BE SUCCESSFUL."
> – Annie, 2nd year graduate student

> "IT'S BEEN REALLY A BIG PART OF FINDING COMMUNITY. I FEEL MORE CONNECTED TO THE INSTITUTION AND THE STUDENTS. AND IT'S BEEN A GOOD WAY TO GET TO KNOW A LOT OF THE FACULTY HERE TOO."
> – Kerstin, 2nd year graduate student

> "I THINK OF MYSELF AS A LIFELONG LEARNER. SO, IT'S BEEN VERY GOOD TO CONTINUE LEARNING NEW THINGS THAT RELATE TO THE COURSE SUBJECT."
> – Kate, 2nd year graduate student

> "IT HELPS YOU COMMUNICATE TO AN AUDIENCE AND ARTICULATE IDEAS SO THEY AREN'T JUST UNDERSTANDABLE TO YOU, BUT TO OTHERS."
> -Spencer, 2nd year graduate student

Most new TAs I have worked with, even those who start out unsure about their ability to be an effective teacher, end up feeling good about their work with students in class. I hope this book will help you get there too.

CHAPTER 1

# Boundaries: Make and Maintain Them

THERE IS MUCH TO DO and think about as you get ready for your first class. Chapter 2 walks you through the steps of preparing for and running your first session. Before tackling those details, we examine a part of teaching that is sometimes overlooked. Specifically, how you define your personal boundaries and how do you communicate this information to students. The ideas described here will help your transition from being a student in your discipline to being a teacher.

On your first day as a TA, you receive a copy of a game called "Boundaries!" You open the box and find three piles of cards labeled "Time," "Sharing," and "Interactions." Shuffling through the cards you see that a scenario involving students is written on each. A card from the Time pile says, "Hold extra office hours by appointment." Another says, "Respond to student e-mails during the weekend." The top card on the Sharing pile says, "Answer questions about my partner or relationships." A card from the Interaction pile says, "Accept a party invitation from a student in my class."

The box also contains a game board. You unfold it on the table and discover that there are two areas labeled, "I'm OK with it" and "I'm not OK with it." At the bottom, there is an area labeled, "Never OK!" Looking closely at the "Never OK!" area you see a list printed there.  The list includes serious-sounding stuff like, "Tolerate aggressive behavior" and "Gossip about students and faculty."

Intrigued, you flip over the lid of the box to learn the rules of the game. You are surprised to find only one instruction: "Place the cards on one side of the board or the other." That's it! Did you miss something? You search the box but, no, that's all the guidance you get.

How do you expect students to interact with you? How much time and energy will you invest in your teaching? How much of yourself will you share with students?

These are some of the boundary-related questions a new TA must consider. There often isn't much guidance when answering these questions. At some institutions, the guidance TAs receive about interacting with students is limited to the policy forbidding romantic or sexual relationships. (These policies and other guidelines set by Title IX in the United States, or similar laws in other countries, are discussed in the Appendix: Guidelines for Teaching.)

Many TAs feel tension between the demands of teaching and honoring their own personal and professional needs. They want to be available for students who need help, but they have limited time and a lot to do. They wish to be flexible in terms of assignments but worry that some students will take advantage of that flexibility. They want the classroom to feel friendly and welcoming but worry that some individuals will not respect the student–teacher relationship. **Most new TAs manage to set appropriate boundaries for interacting with students**. They use their experience in educational settings, their judgment, and advice from peers and mentors to figure out what works for them.

*Back to the board game.* You pull a card from the Interaction pile. It reads, "Give students my personal phone number." You have heard of instructors who were willing to give out their cell number. But to you, a text or call from a student would feel unwelcome. This is a boundary you want to preserve, so you place the card on the I'm-Not-OK-With-It section of the board. The next card reads, "Greet students on campus." It feels natural that you would say "Hi" if you bump into a student from your class. You place that card in the I'm-OK-With-It section. You pull more cards and keep sorting.

The next page shows a sample game board with some common scenarios and interactions placed on one side or the other. The illustration is intended to be a starting point for new TAs—but it might not be quite right for you. You will probably agree with the placement of some cards and disagree with the placement of others.

Over time, as you gain a sense of your own comfort levels with different situations, you can adjust your boundaries—you can move cards! A card placed in I'm-Not-OK-With-It could move into I'm-OK-With-It as your experience working with students grows. Alternatively, you can decide that something you might have originally considered OK is not a good choice for you and move that card to the I'm-Not-OK-With-It area. Defining boundaries is an ongoing process. In your future, you may be handed a card that is not shown on the game board. When new requests for time, sharing, or interaction arise, you will think it through and figure out where to place that card on the board, knowing that you can move it later if it is not working for you.

"JUST MAKING A FEW ADJUSTMENTS."

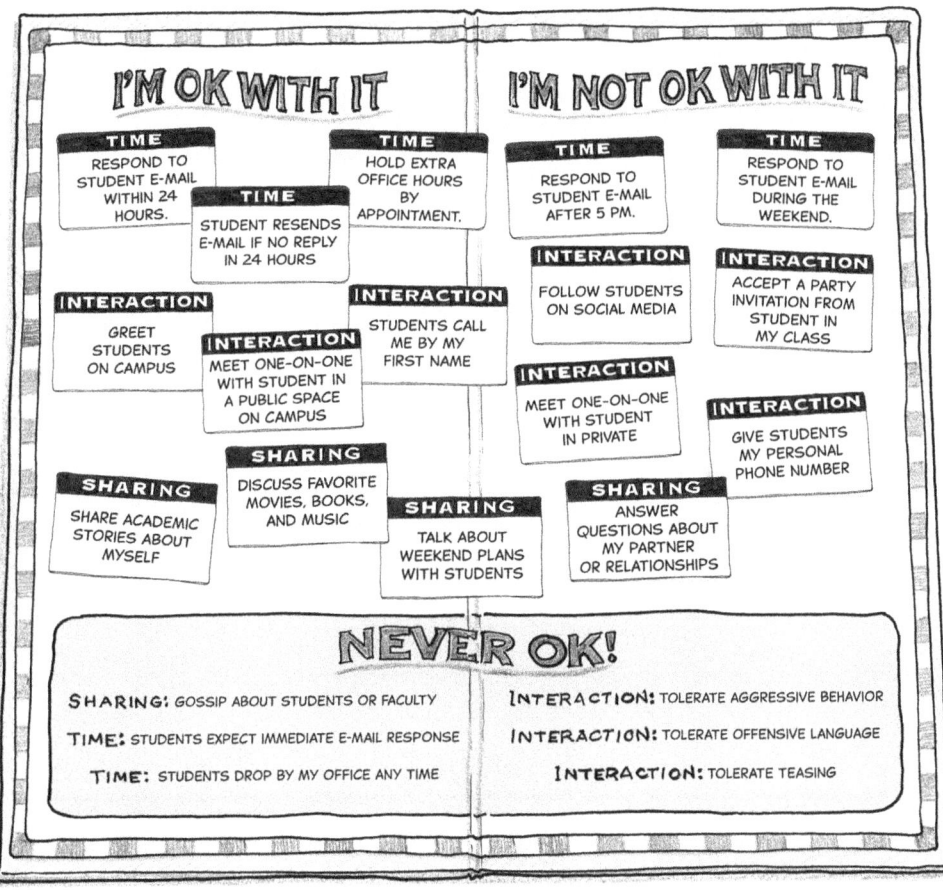

Your boundaries do not have to be fixed and rigid. Perhaps, "Hold extra office hours by appointment." is in your I'm-Not-OK-With-It section. But if a student requests an appointment and they have an appropriate reason for asking, like a work schedule that overlaps with your regular hours, consider saying "yes" if you feel you have time.

The board game analogy is intended to emphasize that defining boundaries is a process involving many small decisions about what you are OK with and that these decisions are personal and editable. The important thing is to make boundaries that honor your physical, mental, and emotional limits. Blurry boundaries lead to stress for a TA and could lead to boundary-pushing by students. Clear boundaries let you maximize the energy and attention you can give to students within those boundaries (Bernstein-Yamashiro and Noam 2013). To quote an experienced TA interviewed for this book, "You want to keep your personal life out of it, and put energy into exposing students to your academic life—your research and your passion for the subject." Clear boundaries can enhance the quality of the mentorship you provide for students.

Before leaving the board game analogy, it is important to take a close look at the part of the board labeled, "Never OK!" This includes requests for time, requests for sharing, and interactions that are never appropriate in an educational setting:

> **NEVER OK!**
>
> **SHARING:** GOSSIP ABOUT STUDENTS OR FACULTY
> **TIME:** STUDENTS EXPECT IMMEDIATE EMAIL RESPONSE
> **TIME:** STUDENTS DROP BY MY OFFICE ANY TIME
> **INTERACTION:** TOLERATE AGGRESSIVE BEHAVIOR
> **INTERACTION:** TOLERATE OFFENSIVE LANGUAGE
> **INTERACTION:** TOLERATE TEASING

Do not engage in or tolerate any of these. At any point during teaching, if you are subjected to aggressive, offensive, or disrespectful student behavior, seek help and advice from your Lab Coordinator, a faculty mentor, or the Department Chairperson. The Title IX office (at U.S. institutions) is another place to turn for help with these situations.

There are far more serious behaviors not mentioned here which are also never OK: physical threats, harassment, racist, or sexist language. If you experience or observe any of these, act immediately. Your safety and well-being are foremost. Tell a supervisor or a mentor. If necessary, contact law enforcement, the Title IX office, or whatever officials are designated to support you at your institution.

## DEFINING BOUNDARIES

The best time to establish boundaries is early in the term and especially on the first day. You will be more relaxed, and you will provide better guidance for students if you start the term with a good idea of what you are OK with and not OK with. Communicating your boundaries begins when students enter the classroom. Some students may not be sure of your role or how to interact with you. When you greet students in a friendly yet formal way you establish the teacher–student relationship. By saying, "Welcome! Please have a seat anywhere" or "Nice to meet you. Please, tell me your name," you are being welcoming while also making it clear that you are in the teacher role.

Once class begins, you will continue to clarify your role and communicate your boundaries. Take a simple example: how you want to be addressed. Students could be confused about this. Do they call you "Professor" or are you on a first-name basis?[1] Let the students know how you want to be addressed on the first day through a clear statement: "Please call me [name or title—whatever you prefer]." This announcement does more than clear up a possible source of confusion. Through this simple announcement, you have begun to establish that you and the students have different roles in the classroom that must be respected. A statement about e-mail communication can also serve this purpose. You might say to the class, "I do my best to respond to e-mail questions

---

[1] In most North American universities, students are accustomed to calling TAs by their first names. In other countries, this level of informality would be considered inappropriate for the classroom (Lambert and Tice 1993). As with many decisions about boundaries, how you ask students to refer to you should be consistent with your personal and cultural preferences.

within 24 hours on weekdays. If you haven't heard back from me in 24 hours, please send a follow-up e-mail." The underlying message is that boundaries exist, and both you and the students have roles and responsibilities.

Chapter 2 describes ways to welcome students to your class. As you think about ways to be welcoming on the first day, remember that you are not there to be friends with the students in your class. TAs must find a balance between being *friendly* with students versus *trying to be their friend*. Go into the class with the mindset that you want to be thought of "as one of them," and you blur boundaries between yourself and the students (Bernstein-Yamashiro and Noam 2013). What if you need to deal with a case of cheating or plagiarism later in the term? What if you need to enforce a class policy regarding late work? Perceptions of friendship complicate these interactions. (This friendly-not-friends advice extends to being friendly online. See Appendix: Guidelines for Teaching for more on appropriate interactions with students in-person or online.)

If a TA is of a similar age to the students in their class, some students may misinterpret roles and boundaries. For example, a student who would not joke with an older faculty member in class might think it is OK to joke around with you. They may wrongly think of you as a peer, rather than as a teacher, simply because you look to be about their age. But you are not their peer, and you may need to address this boundary-crossing student behavior using some of the ideas below.

## MAINTAINING BOUNDARIES

You arrive on campus looking forward to a day with no scheduled teaching or office hours. You have big plans for your thesis research. You have experiments to tend to, papers to read, and a presentation to work on. As you sit down at your desk, you hear "Hello?" and see Michael, a student in your class peeking in the door.

"Hi, Michael. What's up?" You ask.
"I was wondering if I could ask you a few quick questions about my lab report," he answers.

Michael has crossed a boundary by showing up at your office outside of your scheduled hours. As you read about this scenario, it may seem to be an easy call—you would not agree to this request to give up your dedicated research time. But in the moment, with the student standing in front of you, it might not seem so easy. You must either say "No" and maintain your boundary or flex this boundary and alter your plan for the day.

When in-the-moment decision about boundaries comes up, take a "mindful pause" before you respond (Schwartz 2020). Take a breath and ask yourself how this thing you are considering—like a request for your time—contributes or detracts from your goals. Maybe Michael's question will only take a few minutes to answer, but what about the *next* time he shows up? Do you want this one-time interruption to become a regular thing? If you constantly give up your time, you will signal to students that you are always available, detracting from your research and stressing you out. Taking a brief pause to distance yourself from the situation can remind you of the importance of your boundaries and prepare you for the next step in the conversation.

"I can't help you now," you say. "I have office hours tomorrow at 10. I'd be happy to help you then. If you want to e-mail your questions to me, I'll do my best to respond today." Alternatively, if you are willing to provide Michael with a quick answer now, do so and follow up with, "In the future, please contact me before coming to my office or save your questions for office hours."

Here's more advice for having boundary-related conversations (adapted from Wyrick 2022).

**Be direct:** The goal of this conversation is to give clarity to what is and is not OK in your relationship with a student. Clear communication will help you and the student in this situation. Keep it simple. For example:

"I'm not comfortable discussing this topic."

"Your jokes weren't appropriate for class."

"Sometimes your language isn't appropriate for a classroom setting."

"When I ask the class for attention, please stop your conversations to avoid delays."

"It's unfair to the other students in the class for you to continue asking for extensions to assignments."

"I'm upholding the class policies [on absences, late work, etc.]."

**Do not be apologetic:** You do not need to apologize when you maintain boundaries. There is no need to start these conversations with statements like, "I'm probably being too sensitive, but…" or "I know some instructors are OK with this, but… ." Qualifying boundary-defining statements in this way makes the message less clear.

**Avoid calling the student out:** Conversations with students about boundaries should be conducted privately when possible or even electronically through e-mail. If it is necessary to address a boundary during class, try to do it discreetly. Your communication about boundaries is more likely to be received by the student if their peers are not privy to the conversation.

## BIAS AND BOUNDARIES

It would be wrong not to acknowledge that issues around boundaries may affect some TAs differently than others based on age, gender, race, or ethnicity. As you read this section, remember that the negative experiences described here are *not* universal or inevitable. But these experiences are common enough to make them essential topics for TAs to think about as they start teaching.

TAs who appear young, who present as female, and/or are from underrepresented groups may find that some students test their boundaries in ways that those students would not test TAs who are older, male, or white. Boundary challenges sometimes take the form of students being "pushy" or more likely to argue about grades and class policies. They may take the form of students acting too informally toward their TA—inappropriate joking, teasing, or just generally not respecting the student–teacher relationship. In another form of boundary-crossing, because of the TA's identity, students may see the TA as a person to whom they can disclose emotional information about themselves. You may feel honored that a student feels safe sharing their emotions with you, but listening and empathizing is not without time and energy costs to you (Rockquemore 2015).

Mara is a 21-year-old Biology graduate student who identifies as female. Here she talks about her experience:

> **Interviewer:** What aspect of teaching has been the most challenging for you?
>
> **Mara:** Being a female TA is different and more challenging in some ways than being a male TA, especially being young. I had a class that was 18 boys and four girls and felt I never really had authority in that classroom. I think they felt entitled to things or were more comfortable pushing boundaries. Like they would ask me to make up [assignments], or they would ask for other accommodations. I just felt like they were asking me for a lot of things—a lot of favors—that they maybe wouldn't ask another instructor.
>
> **Interviewer:** How does that experience influence your teaching now?
>
> **Mara:** Looking back and having talked to [my research advisor], there are things I am doing differently, like I set a lot firmer boundaries in the beginning and start off a lot more strict. I wanted to be nice, and I wanted them to like me. I think I was pretty easygoing, but that might have [led to] the boundary-pushing later.

Mara's experiences are not unique. Student bias around gender, race, and ethnicity in higher education is well-documented (Chávez and Mitchell 2019; Kreitzer and Sweet-Cushman 2021; Llorens 2021). One study looking specifically at bias in teaching evaluations of TAs provided empirical evidence for the gender bias that TAs like Mara have experienced (Khazan et al. 2019). Dr. Kerry Ann Rockquemore, in an essay about boundaries, addresses the extra boundary-related challenges that some college teachers face because of their identity: "I'm not saying it's fair that you may have to frequently defend your boundaries. I'm saying that's real and it's why you have to be extra clear about where your boundaries lie and skillfully push back when students cross them" (Rockquemore 2015). Mara's advice—formed through experience and consultation with a mentor—to err on the side of setting firm boundaries early in the term is wise.

### Want to Know More?

Rockquemore KA. 2015. *How to listen less.* https://www.insidehighered.com/advice/2015/11/04/setting-boundaries-when-it-comes-students-emotional-disclosures-essay

> The author focuses on ways to handle situations in which students cross boundaries related to emotional disclosure and provides advice about discussing boundaries with students.

Schwartz HL. 2020. Role clarity: how faculty can map their own boundaries. *NEA Higher Education Advocate,* **2020**: no. 08.

> A concise guide to how identity, power, and institutional norms can shape student/teacher relationships. The author shares ideas for setting, maintaining, and adjusting boundaries in-person and online.

CHAPTER 2

# Your First Class

WHEN YOU BUY A NEW device, it usually comes with a thick manual. Most of us toss that aside in favor of the one-page, illustrated "QuickStart" guide. Think of this chapter as the QuickStart guide to teaching a lab or discussion section. Almost everything in this chapter will be discussed in more detail later. Read ahead if you can, but this chapter gives the minimum to get your class off to a good start. Starting your class off right comes down to three things: (1) be prepared, (2) be welcoming, and (3) give the students a chance to talk.

Your first session will *not* be perfect. It is unrealistic to expect that you will anticipate every question, manage time expertly, and perfectly articulate every bit of information. You will try to do your best, but there will be things you will want to improve on. (At the end of this chapter, there is some advice for handling any missteps from your first session.)

### BE PREPARED

How would you answer if a student asks you, "How should I prepare for class?" For a lab class, you might recommend reading the background and procedures in the lab manual and making a simple flowchart of what you will be doing. For a discussion section, your advice might be to review the lecture notes, attempt all the homework problems, do the relevant readings, and keep a list of ideas or concepts that you have questions about. (Wouldn't it be great if students did all these things prior to class?)

Consider how this pre-class preparation needs to be different for you as the teacher. Ideally, you will be better-prepared than your best-prepared student. In addition to everything described above that a well-prepared student should do, you also need to consider the timing of the session and the order of activities. You must plan what you are going to say to the students and what you are going to ask them to do. Finally, you'll do your best to anticipate questions and how you might respond.

That is a lot to think about. Plan to dedicate a few hours to your preparation for your first class—more if you are feeling nervous. As you gain teaching experience, the time needed to prepare for class will shorten. For your very first session, give yourself plenty of time to get ready. There will be a lot going on in your head and in the room when the class starts, so think through it ahead of time. (Students trying to add or "crash" the class can make your start-of-class plans more complicated. See Box 1 for advice about dealing with the common first-day issue of crashers.)

It is likely that you will not be starting from zero when you prepare. For most classes, especially those with multiple lab or discussion sections, there will be some sort of support for you, but this will differ from class to class and from institution to institution. Support could be in the form of weekly coordination meetings between the TAs and a faculty instructor. Support could be in the form of a written TA guide. These forms of support are helpful tools, but they are just a starting point. Before your class starts, your preparation should include **writing out your own set of notes** that will be useful to you during the session. This is your Personal Teaching Document.

The next pages show real examples of notes for an introductory laboratory class. First is the outline provided by the lab coordinator. The outline is helpful, detailed, thoughtful, all-inclusive, and—sorry!—mostly useless when you are up in front of the class. The type is too small to reference quickly as you speak, write on the board, or move around the

---

**Instructor Guide**
**Membrane Diffusion: Interaction between Cells and Environment**

*Remind students that there will be no report required for this lab, but that there will be a quiz at the start of our next meeting that will cover the results/conclusions from this lab activity.

**I. Diffusion and Osmosis (~40 minutes – while waiting for dialysis, have students complete "Water Molecules in Motion" activity)**

1) Explain the difference between the cellulose membrane that will be used in class for dialysis and the cell membrane in terms of their composition and their permeability (Note: dialysis tubing = mesh of cellulose fibers, cell membrane = phospholipid bilayer, integral membrane proteins). Consider making a table and asking the students to help you fill it in.

2) Remind students of the three solutions being tested and the three tests being performed on each solution (total of nine tests)

   a. 3 tubes of each: Stock solution (SS) / Bath water before exposure to the dialysis tubing (BB)/ Bath water after exposure (BA)
   b. 3 tests for each solution: for salt (SA) / for starch (ST) / for sugars (SU)

3) Demonstrate how to prepare the dialysis bag (synchronize the dialysis step!)

   a. Wet pre-cut tubing for ~ 1 minute in glass bowl in DI water at bench and pry open by rubbing between fingers
   b. Twist one end 1 inch from bottom and bend the end at the twist back towards the rest of the tubing
   c. Use string to tie the folded and twisted end using standard box knot (very tightly)
   d. Pipet into tubing ~8-10 mls of stock and twist/fold/tie other end with as few air bubbles as possible.

## BOX 1. CRASHERS

It is possible that your first-day plan could be interrupted by students trying to add or "crash" the class. These students will usually do one of two things: They will walk straight up to you and ask very directly, "Can I add your class?" or they will come in quietly, take a seat, and wait. Both approaches can complicate your start-of-class plan so it is worth thinking through both scenarios ahead of time. Be sympathetic and help these students if you can, but do not let them derail your start-of-class plans. It is OK to delay the start of class by a few minutes to sort out potential adders.

If you are teaching a section of a large-enrollment class, it is likely that someone else (probably the lecture instructor) is responsible for adding students. Alternatively, you may have the authority to add students if space allows. Know these policies and do not make promises. You may tell a student "There's one seat open. You're in!" only to find out that the lecture instructor gave someone else the seat an hour earlier. Even if you are willing to have a few extras in the classroom to accommodate students desperate to enroll, your institution may set limits on room capacity, and students you promised a spot could get denied later when they try to add the class officially. Here are the steps to take as you deal with this common first-day-of-class situation:

- **Step 1:** If a student is standing in front of you asking to be added, politely ask them to wait off to the side while you finish your setup (plug in your computer, write key information on the board, etc.). This gives all the enrolled students time to show up.

- **Step 2:** Announce, "If anyone is trying to add the class could you please come up to see me." If there are nonenrolled students who have just come in and taken a seat, this will help you identify them. Now you know exactly how many potential adders you are dealing with.

- **Step 3:** If you have a small group at the front of the class, you can quietly explain the enrollment procedure *once* to the group. "You need to contact the lecture instructor. Here's her e-mail address." or "Add your name and e-mail to this wait list and you will be contacted if a spot opens." Knowing the class policy is key.

If you have the authority to add students to your section, it can get *very tricky* if there is one spot and, say, five students trying to get in. Should you make a random choice right there on the spot as you face five anxious students who all have valid reasons for wanting the class? There may be an existing wait list for the class that provides a neutral way to prioritize the students. If there is no wait list, find a way to delay the decision. You could ask all the students to contact you via e-mail explaining their need for the class, allowing you to make a need-based decision on your own time. It is not a crisis if a student adds to the class after the first session. With some coordination that student could sit in on another section later in the week or meet with you during your first- or second-week office hours.

room. This page shows the same information rewritten in a form that is useful during the actual class. Besides being written large enough to see while standing and speaking, these handwritten pages have the following benefits: (1) everything you will write or draw on the board is written or drawn in the notes, (2) every important statement you will make is easy to see, and (3) important points about lab procedures are highlighted so you do not have to worry about remembering them. Creating this Personal Teaching Document is an essential part of your preparation.

*Excerpt from a typical instructor guide:* An outline like the

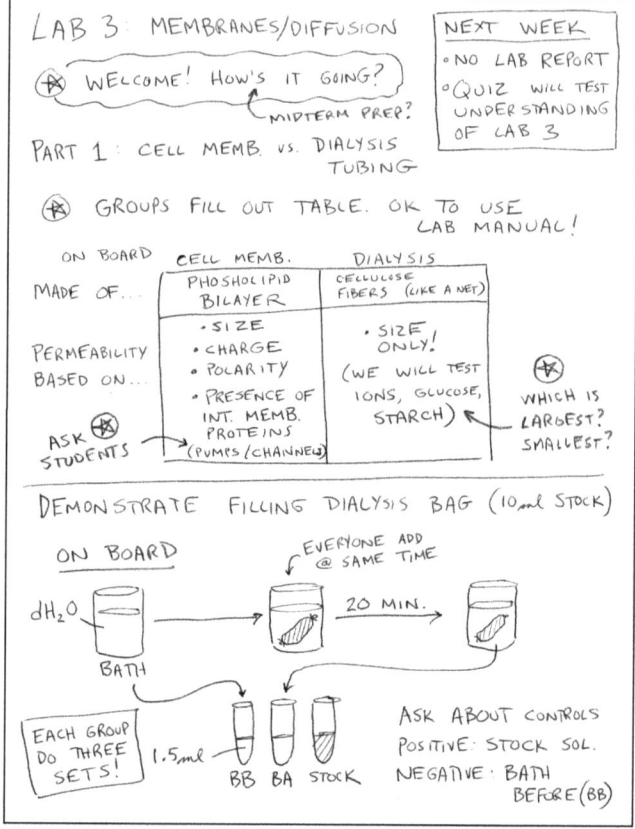

one on page 14 might be provided by your course coordinator or someone with experience with the class. There is detail on content, timing, and procedures. However, the format is not easy to access while teaching. It is also missing key information, like what you are going to say and what you will write on the board.

***One page from a Personal Teaching Document:*** This example corresponds to the outline in the previous figure. This handwritten page is useful while teaching. The writing is large enough that you can see it while speaking and moving. Everything you plan to say or write on the board is included—maybe not every word, but short prompts that remind you of your plan. Questions you will ask the students are marked with stars. Consider using multiple colors to distinguish the content you will provide from the content you will ask the students to provide. Each page of the typed outline (see page 14) will generate one or two handwritten pages.

***Personal Teaching Document from slides:*** If you plan to use PowerPoint or a similar program to project slides while teaching, it is important to add notes about what you will say in the classroom. Print out your presentation (two slides per page will give you plenty of room to write) and add notes as you plan and practice your lecture. Is there information you need to emphasize for the students? Is there an opportunity to have the students talk or answer questions? Prompts written on your notes help you remember these details (see page 17).

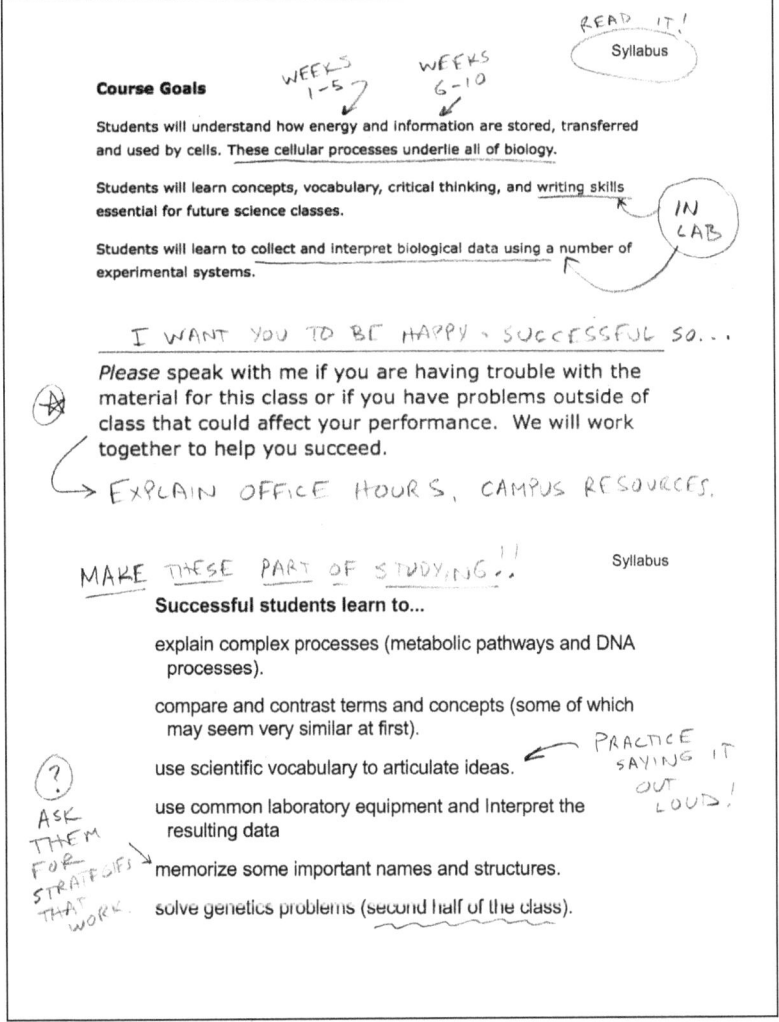

At some institutions, TAs are required to observe a more experienced TA prior to teaching their own section. If this is not the norm at your institution, ask another TA if you can sit in on their class. Every minute you spend observing another teacher is time you spend preparing for your own class. Observe the class from the perspective of a teacher and try to identify the reason behind the instructor's actions. Take notes on timing and interactions with

students. What questions does the instructor ask that get good responses? It's OK to "steal" great teaching ideas and use them. If you are observing a laboratory class, use the time to get some hands-on experience with the equipment. Try out the lab activities yourself. Notice what tasks are easy or hard for the students and anticipate questions. This is time well spent.

Familiarity with the physical space is another important part of preparing for your first class. Avoid any last-minute distractions from the more important things you'll do on the first day. At the very least, locate the classroom. This may be easier said than done if you are a first-year graduate student on a new campus. If there is an opportunity, early in the morning or in the evening, go in the classroom and explore. Do you need to bring your own whiteboard pens or chalk? Will your laptop connect to the projector? Is an adapter needed? If the room is busy all day, it is fine to explore during the transition between classes. If there is another  instructor setting up say, "I'm teaching in here later. Do you mind if I look around?" They almost certainly will not mind and might even give you some helpful tips if they have experience with the room.

It is normal to be nervous before your first class. Nothing I can write here will eliminate all anxiety. However, preparation and practice will help. A good strategy is to plan your first ten minutes thoroughly and then practice it out loud. Thinking about what you are going to say and making words come out of your mouth are not the same. (For advice about speaking in front of the class and writing on the board, see Box 2.) Before my first-ever class, I found an empty classroom late at night and gave the lecture to no one. It felt a little weird—and I was glad no one came in while I was doing it—but it helped me get comfort-  able. I know TAs who find it very helpful to do a practice walk-through in the empty lab, pointing out materials and equipment as if students were there. If you are nervous, practice it out loud to a roommate, another TA, or to an empty room.

## BOX 2. WRITING AND SPEAKING IN CLASS

**Writing on the Board**

Writing on the board is less common than it used to be. Most classrooms have a digital projector making it very easy to show "slides" that you have prefilled with information. Having everything on the slides can be helpful, especially for a new teacher. You can think everything through ahead of time: what information to include, what order to present it, etc. Having everything in the slideshow means you can rely less on your notes and memory. The decision of whether to use slides may not be up to you. If you are part of a large multisection course, there may be a standard set of slides that every TA is expected to use.

Even if you do you use slides, the chalkboard or whiteboard is still a useful tool in teaching and it is worth thinking about how best to incorporate it. There may be times when writing on the board is necessary or preferred. For example, having a written outline of the session on the board can help you and students stay organized. Equations or vocabulary that you want the students to access throughout the session can be written on the board as a reference. Even if you have a slide presentation, students may ask for clarification or more detail that requires writing or drawing.

We write on paper using fine movements of our hands and wrists. If you try to write on a chalkboard or whiteboard the way you write on paper, it will be slow and difficult to read, and your hand will get tired. Writing on the board is more like drawing than writing. Take a tip from artists—when you write on the board, keep your hand and wrist mostly still and move from your shoulder. This takes a little practice, and your board writing will not look much like your handwriting; however, you will find this lets you write faster and more legibly.

If you want a hybrid approach where you can show slides and images and still incorporate writing, a tablet with a stylus that lets you write on the screen is a great option. Notetaking apps for these devices make it easy to flip through your slide presentation and add blank pages for writing and drawing as needed. Some advantages of this approach are that you can flip back to any image or handwritten page easily, and you have a digital copy of the entire presentation including what you wrote and drew. This digital record can be used to augment your personal teaching notes from the session and be shared with students.

**Public Speaking**

Everyone develops their own style for speaking in the classroom. Your style will develop from your own personality with inspiration from your former teachers. Advice about public speaking is easy to give but hard to follow. Try not to go too fast or too slow. Try not to speak too loudly or too softly. Like anything with teaching, your speaking speed or volume will not be perfect, especially not at first. You will make changes and improve if needed. (Regarding speed, most new TAs realize that in their first session they talked really fast. Like them, you may need to remind yourself to slow down, breathe, and give students a chance to catch up with note-taking or to ask questions.)

The best way to evaluate and improve your public speaking is to get feedback. At some point during your first session, find a moment to ask a student in the back half of the room, "Could you hear me all right?" or ask, "How was the speed? Too fast? About right?" Another simple way to get feedback on your speaking is to give your phone to a student and ask them to record a video while you are speaking. One minute of video will be packed with information you can use to improve your public speaking. Are you mostly facing the class when you talk, or do you have your back to them? Do you look somewhat happy or are you scowling? If you are like most of us, you will find watching this video a bit cringey—and you do not need to share it with anyone! But seeing yourself moving and speaking from the student's perspective will help you identify many ways to improve your classroom speaking. (If you do not like the idea of handing your phone to a student to record a video, you can still get useful information from an audio recording using a dictation app. Just hit "record" and set your phone down when you start teaching.)

## BE WELCOMING

Students should feel welcome in your classroom. You are the authority in the room and your words and actions will set the tone for all the interactions that follow. A safe, friendly environment is good for learning, so get off to a good start by welcoming your students to class. As described in Chapter 1, as you welcome students to the class, you will also be establishing your role as the teacher and beginning to communicate your boundaries.

You may be required to take attendance on the first day. Rather than call out names from the front of the class, consider walking around the room, greeting each student, asking their name, and making notes about pronunciation. Besides giving you an opportunity to have a one-on-one moment with each student, there are some advantages to taking attendance this way from a privacy standpoint. Some students may have opted for privacy around their personal information as part of the Family Educational Rights and Policies Act or FERPA. These students may not want their full names to be read out loud in the classroom. (Read more about FERPA and how it affects your classroom in the Appendix: Guidelines for Teaching.) Once you have started the class, do not bring it to a halt to do attendance for a late student. If students come in late, as often happens on the first day, just announce, "If I didn't check in with you before class, please see me (after class, at the break, etc.)." The student will appreciate not becoming the center of attention simply because they had a little trouble locating the classroom.

Avoid the temptation to start with "going over the syllabus." Time spent introducing yourself, welcoming your students, and having your students interact with each other is important because it sets the tone for the learning to come. An instructor who immediately launches into the details of the course (schedule, grading, etc.) sends a message that those elements, not student learning and student experience, are most important to them. Plan to take a few minutes to set a positive tone.

***Introduce yourself:*** One of the best ways to welcome students is to share something from your own undergraduate experience. Did you take a similar course to the one you are teaching? Have you faced educational challenges in a course like this one, and if so, how did you overcome them? Being open about your experiences as a student will start to build trust in your classroom.

***Tell students how you want to be addressed:*** Students might be unsure how to address you. Do they call you "Professor," or do they call you by your first name? How you wish to be addressed is your decision. (Read more about this and similar decisions in Chapter 1.) Letting students know how you wish to be addressed does more than clear up confusion—it signals to students that you have different roles in the class, even if you are on a first-name basis.

***Introduce the content:*** What is the course about and why does it matter? Does it lay the groundwork for success in advanced classes? What parts of the class are you especially excited to share with the students? How is the content relevant to students' lives? (This last question is especially important if you are teaching a non-majors course like a biology course for future architects. These students might not have thought about the relevance of the subject.)

***Say "I'm glad you are here:"*** Let your students know that you appreciate them taking the class and that you are excited to help them learn. (See Box 3 for more examples of welcoming language for the first day of class.)

---

**BOX 3. INCLUSIVE LANGUAGE FOR THE FIRST DAY**

Sending a positive, welcoming message on your first day of class is important. This message should also be inclusive so that all students, regardless of their backgrounds, feel welcome (see much more in Chapter 5: Inclusivity). Here are some suggestions for inclusive, welcoming language you can use on your first day (adapted from Blunt et al. 2018):

- "**You belong in this class**. You are smart, hardworking, and capable of learning complicated material."
- "**I want to be a mentor, not just a teacher.** Tell me when you are struggling with the class material or if your identity affects your experience in the class. We will work together to help you succeed."
- "**Many students struggle with imposter feelings**. They feel that they don't really deserve to be here and will eventually be 'exposed' as a fraud. You are not an imposter and you deserve to be here. I'd be happy to talk with you about imposter feelings any time."
- "**I want this classroom to be a good environment for everyone** regardless of gender, sexual identity, race, ethnicity, or religion. Please feel comfortable coming to me if you ever find yourself feeling unwelcome for any reason."

## GIVE STUDENTS A CHANCE TO TALK

Picture your ideal class session. It is likely that your mental picture includes lots of *productive* talking: students discussing ideas with each other and asking and answering questions. This is a good goal, but it will not happen on its own. Set the expectation that students will interact in class by giving them chances to do so right from the beginning. A class session dominated by the teacher talking sends a message to students that they are information receivers, and their active participation is not valued—maybe even discouraged.

If your goal is a classroom where students discuss ideas and opinions, give them at least one opportunity to do so near the start of your first session. This sends the message that you expect and value in-class discussion. Talk can be focused on the content students will learn or on building class community with "get-to-know-you" topics. For example:

> *"Introduce yourself, say your major, and share why you chose that major."* This prompts students to share something about what they enjoy or what motivates them.

> *"Introduce yourself and share your best studying tips (for any class)."* A benefit here is that as the study tips are shared with the class you can "second" those that are particularly important for this class.

> *"Introduce yourself and share one interesting thing you learned in the last year about [biology, astronomy, psychology, etc., whatever the subject]."* This prompts students to think about their own interests related to the subject of the course.

If your class is small, you can ask each student to respond. In a larger class, or if time is limited, ask the students to share their responses with three or four students around them. Any time you ask students to discuss something in small groups, watch for students who are not in a group and be prepared to help them join a nearby group. Alternatively, you become their partner for the activity. With a bigger class you might say, "Thank you for talking with each other. We do not have time to hear all the ideas, but would anyone like to share something interesting that they learned from another student?"

At some point in your own education, you have probably heard an instructor respond to a student question with, "It's all in the syllabus." Avoid this response! Yes, the student's question is most likely answered in the syllabus, and, yes, students are responsible for reading and understanding the course policies. However, put these considerations on hold to send an important message to the class—that you value student questions. You may restate something you just said or give information that can be easily found in the syllabus, but answering will only take a short time and lets the students know that their questions are important.

If your first-day agenda is so packed with content that you do not feel you have time to answer questions on course policies, think critically about what you are covering and try to free up some time. Alternatively, if you think you will be very short on time on the first day, send the students an e-mail in the days before the class: "I'm looking forward to meeting you. Our first session will be action-packed and we won't have unlimited time to answer questions about the course policies. Please read the syllabus (attached to this e-mail or linked on the course website) prior to class so we can get off to an efficient start." Sending an e-mail like this does not mean you skip talking about course policies or answering student questions about those policies. It sends a polite message that time in class is precious and you will not have *unlimited* time to answer questions.

Once you are into your presentation on course policies or content, make sure you stop to ask for questions. Consider saying, "I've gone over lots of details. Is there anything that needs more explanation?" (Phrasing the question this way invites more responses. See Chapter 6 for more tips on asking questions that get a response.) Alternatively, create an opportunity for students to talk that allows them to see each other as sources of help and support. For example, after going over the syllabus and course policies you could say, "In groups of two or three, take one minute to discuss any questions you have about the course." After one minute, ask the class, "Do any groups have questions for me?" This is a version of the Think–Pair–Share strategy described in Chapter 7. Notice that by following these steps, you are not asking for an *individual* to raise their hand and say, "I'm confused about ... ." Instead, the questions are coming from a *group* where everyone was unsure about a particular detail. This approach reduces peer pressure associated with asking a question on the first day.

## COMMON FIRST-DAY CONCERNS

***What if the students don't participate?*** Be prepared for the class to feel a little unfriendly at first. After all, they are strangers to you and mostly to each other. Do not expect them to smile. (Honestly, it would be weird if they *were* all smiling at you.) Despite an outward appearance of disinterest or unfriendliness, the class is largely made up of students who are happy to be there. Give them the opportunity to participate and show them that you expect them to participate. Keep in mind that they may need a little time to warm up. If you have time before your first class look through Chapter 6 for ideas about how to encourage students to get involved. If you ask a question to the group and you get no response, there could be multiple possible explanations. Maybe you did not ask the question in a way the students understood. Say, "Let me ask that another way," and

STUDENTS WILL NOT LOOK LIKE THIS ON THE FIRST DAY.

try again. Alternatively, the students might be feeling peer pressure on the first day and do not want to appear too eager to participate. Ask them to talk over the answer with students around them for a minute, and then ask a group to answer. This lowers the participation barrier. Maybe you just need to wait a little longer after asking a question and hands will go up. If you want a class in which students are active participants, it is important for them to participate on the first day, even if it takes a little more time and effort on your part to get the conversation started.

***What if someone asks me a question and I don't know the answer?*** Will this happen? Most likely, yes. It is unrealistic for you to anticipate every question about course policies or about the content you are teaching. It is fine to say, "I don't know the answer to that, but I'll find out."

New teachers can be very self-critical. Even if your first session goes mostly well, you might find yourself focusing on the bit of information you forgot to convey or the question you were unable to answer. That's what e-mail is for! Consider sending something like this to the students:

> *"It was great to meet everyone in class today. There were a few questions that I wasn't prepared to answer during our session, so let me answer those now. [Restate questions and provide answers.]"*

Beyond providing useful information about the course, an e-mail like this can help you get beyond the feeling that you "messed up." It also shows the students that you care about doing a good job. Imagine you were a student in the class, and you got an e-mail like this shortly after the first meeting. Wouldn't this communication give you a positive impression of your TA?

***What if the students don't respect me?*** You were once in the students' position. How did you view your instructor? You probably assumed your instructor was selected to teach the class for a reason—experience, advanced content knowledge, etc. —and you gave that person your respect. Your position as TA comes with built-in respect from the students whether you feel you deserve it or not—and you do deserve it! If you do your best to be prepared and welcoming, you will keep that respect. College instructors experience imposter feelings, and concerns about losing student respect ties into those feelings. Maybe you are not an expert on every aspect of the content you will teach ("I haven't thought about photosynthesis since my first year as an undergrad!"), but you have shown that you have the academic skills necessary to excel in your discipline; you know how to learn. It is this set of skills that ensures you will make valuable contributions to the students' education.

The respect the students have for you is resilient! It will not be shaken by a single "bad" session or mistake in the classroom. However, the built-in respect that comes with your position *can* erode over time if you are unprofessional: You are consistently underprepared, you are rude or dismissive to students, or you do not follow through on commitments like grading or office hours. If you strive to be prepared and professional, to be responsive to students, and try your best, the students will respect you.

CHAPTER 3

# Reflection: Capture Your Experience

CONGRATULATIONS ON FINISHING YOUR FIRST class. You did it! Before judging the experience, let us acknowledge that teaching is hard. Teaching is one of the most complex cognitive tasks you can engage in—one in which you use many parts of your brain at once. Teachers simultaneously apply creative, administrative, logical, and interpersonal skills—all while standing and speaking in front of a group. You should be proud that you have boldly stepped across the line from being a college student to being a college teacher.

Given the mental challenges of teaching, it is not surprising that new teachers can feel very tired after class. Feeling wiped out is normal and a sign that you are doing it right. You will not always feel this tired after teaching. As you gain experience, teaching will not be as physically taxing.

Like most new teachers, you probably wish you had done some things differently. This is normal, and you will improve with practice. No one is born knowing how to teach a lab or discussion section and you are still at the start of the process.

New teachers tend to be very self-critical. They focus on small negatives: a question for which they did not have an answer, a bit of information they forgot to mention, or an awkward student interaction. Even if your class went mostly well, you might find yourself dwelling on a small thing that you wish you had done differently. You can use the recommendations in this chapter to move past this feeling in a positive way by recording your ideas of what to change and how to improve. But try not to let something relatively small cloud your impression of how you did. Acknowledge what went well in addition to thinking about what needs improvement. If you do that, you will likely identify plenty of positive moments from your teaching. (Sending an e-mail to clear up a point or to provide additional information could help you get over feelings that you made a mistake. See the previous chapter for wording you can use if you choose to send an e-mail like this to the class.)

It is very tempting to shift your thinking to next week's class, but there is something important to do first—something that will improve your teaching and make your life easier in the future. Before moving on, **reflect** on this week's teaching and collect your thoughts—briefly—in writing. The best teachers are reflective. They look at what they did in their class, think about why they did it that way, and how successful it was for the students and for themselves. Reflection is a key to improving as a teacher. Graduate students who engage in regular reflection are more likely to see themselves as effective teachers and are more willing to explore new ways of teaching (Schussler et al. 2008).

A major benefit of writing a reflection is that capturing your experience will save you time in the future, allowing you to recoup the time investment you made preparing to teach this week. Now that you have taught the class, you probably have ideas for small fixes and maybe for some bigger changes. Record those ideas in writing while the experience is still fresh in your mind. No matter how vivid the memory is right now, you will not remember all the little details of your session in the future. These notes will ensure that you can build on your experience.

Was there a question you asked the class that got a good response? Write the question down so you do not have to expend any energy thinking it up again. Was there a topic the students were confused about and how did you address it—or wished you had addressed it? Make a note of the question and your response because it will probably come up again in the future. Where are the pH test strips—that you spent five minutes looking for—kept? *Write it down* now. (They are in the cabinet under the fume hood, by the way.)

Take ten minutes to write the teaching reflection. You will find that these 10 minutes are as valuable as any of the time you spent planning to teach this week. Use pen and paper or keep a digital record. Consider adding a page or two onto the end of your teaching notes for that lab or session. If you prefer a digital reflection, start a document on your computer or on the cloud. The important thing is that you do it and that you have a way to keep track of it for the future.

"MIRROR, MIRROR ON THE WALL, GROUPS OF TWELVE ARE TOO BIG FOR A PRODUCTIVE DISCUSSION."

The audience for the reflection is you! You are writing to your future self about what went well, what needs improvement, and, if possible, ideas for how to make those improvements. Include anything you think will be helpful to your future self as you prepare to teach this class again or a different class. Consider including the following information:

*What went well in your teaching this week?* Make notes about things that happened in the classroom that were successful for you and for the students. Maybe you asked a question that got a good response and sparked a discussion. Write that question down in as much detail as possible. You will not have to remember it the next term if you reteach the same class. Was the timing for each activity correct? Indicate about how many minutes each activity took. Did you explain something in a way that seemed to "hit home" with some students? Record these notes here.

*What from this week needs improvement, and how can it be improved?* If you have specific ideas about what you will do differently next time, record those here. These can be small details ("I need to practice catching amoebas from the jar so I can help students do it.") or bigger changes to your approach to teaching the material ("Rather than lecture about protein structure, ask students to fill in the details on the table on page 10 of the lab manual. Missed opportunity to let them talk and review.") This part of the reflection should be a list of actions you can incorporate into future teaching.

*Notes about technical or management issues associated with this session and how to address them in the future.* Was there traffic jam around a shared piece of equipment? Did you rush through the introduction only to have class ended 30 minutes early? Would the activity have been better with groups of two or three rather than groups of five or six? It only takes a few seconds of writing now to avoid small inefficiencies in the future.

*What questions did students ask?*

*Anything else* that could be useful in the future.

The more detail you include in the written reflection, the more valuable it will be later. For example, reflecting after teaching a lab section, you might write, "Some groups barely had time to finish because they had to wait for the buffer to thaw." This note is helpful, but why not use the reflection to troubleshoot and plan. Follow up with, "Next time, write 'Get buffer out to thaw!' on the board and check with each group when they start working." The lab will go more smoothly, and you will feel more relaxed—all thanks to a few minutes of thinking and writing now.

There are no rules to writing a teaching reflection other than to do it. Your future teacher-self will thank you.

## USING THE REFLECTION

TAs Sonja and Tara both taught lab sections of *Introduction to Botany* in the Fall and are preparing for the Week 1 lab of the Spring semester. Both TAs have their teaching notes from last semester, but only Tara has a short, half-page teaching reflection. Tara wrote the reflection while sitting in the empty lab room immediately after she finished teaching the Week 1 lab in the Fall. She stapled the reflection to the back of her teaching notes and tucked all the pages under the cover of her copy of the lab manual.

Both Sonja and Tara have "Icebreaker Activity" written in their teaching notes. Sonja does not remember the exact question prompt she used last semester and starts searching the web for college icebreakers. Tara's reflection includes the prompt and some details about how it went:

> "Icebreaker: Groups of 5 students. 5 minutes (good timing). Asked students to introduce themselves and share an example of a plant that's important in their family or cultural identity. I shared memory of cooking and eating okra with my grandmother."

After reading the reflection, Tara requires no additional preparation to get this activity off to a smooth start.

The lab has started. Sonja notices some students working with large leaves collected from the trees around the building. She remembers something from last semester. "Hey everyone!" she shouts to get the students' attention, "This works best with young leaves so, if you're working with a leaf that's bigger than a credit card, you should start over!" About half the students head back outside to collect a new leaf. In Tara's Teaching Reflection from Fall, she had written:

> "Young leaves work better. Show them the size BEFORE they go out collecting."

---

**HOW DO YOU KEEP TRACK OF REFLECTIONS FOR YOUR CLASSES?**

> "I HAVE A NOTEBOOK, AND EVERYTHING FROM LAST TERM IS WRITTEN IN BLACK. EVERYTHING FROM *THIS* TERM IS WRITTEN IN RED. THAT'S A VERY EASY SYSTEM, BECAUSE IT DOESN'T CREATE A WHOLE OTHER SET OF MATERIALS. IT'S LAYERED SO I CAN REMEMBER, 'THIS WAS HARD FOR ME.' AND I'VE GOT A NOTE ABOUT HOW I DISCOVERED TO DO IT. THAT MAKES IT A LITTLE EASIER. YOU CAN RECREATE THE THOUGHT PROCESSES AND SEE YOUR EVOLUTION IN TEACHING THAT TOPIC."
>
> -Maria, 1st Year graduate student

With this reminder, Tara grabbed an appropriately sized leaf on her way into class and held it up during her introduction saying, "Young leaves work best. Look for something this size."

These stories illustrate how the short time required to write a reflection, done soon after teaching, can make things easier for you and the students in the future. It is not realistic to think, as Sonja did, that you will remember small details from the previous term. Sonja did not do a bad job in the classroom, but she wasted time and energy before and during the class.

## REFLECTION TEMPLATE

In the back of this book, you will find templates for writing your teaching reflection. (Templates can also be downloaded at TheTAsGuide.com.) There is one template for each week of a 15-week semester. A sample reflection for an introductory biology lab that uses this template is shown on the next page. Use the template if it is helpful. Alternatively, create your own template or create a file on your computer and update it after each class session. Regardless of the format, make sure you write a reflection and that you have an easy way to keep track of it for the future.

## A SAMPLE REFLECTION

| |
|---|
| **Date or Week of the Term:** Week 2 |
| **Name of Lab or Topic of Discussion:** Membranes, Diffusion, Osmosis |

| |
|---|
| What went well this week? |
| 10 min quiz at the start of lab. Good length…only a few students took the whole time. I worked on memorizing names during the quiz. My Intro was a good length (under 10 min) and students got to work right away. |
| Plenty of time to do this lab. Most groups finished with about 15 min to spare. I encouraged them to start drafting the lab report rather than leave early. Some groups did! |
| Having lab groups share/compare data before graphing. 6 groups of 4, 10 min. Did this at the 2-hour mark. (One lab group had to repeat an experiment so they did that during the data share.) |

| |
|---|
| What from this week needs improvement? |
| I lectured about membrane structure, but it would be better as a discussion. (Ask each table to draw membrane, label hydrophobic, hydrophilic regions, etc. Use whiteboards?) |
| I didn't know how to get the lab computers to draw a best-fit line on the graph (an old version of Excel). We figured it out but it took a while. (Chart -> Scatter chart -> Add Trendline) |

| |
|---|
| What technical or management issues arose in this session? How will they be addressed? |
| Traffic jam at station 1. Groups didn't realize that they could do stations 1, 2, and 3 in any order. I sorted it out but remember to mention it in the intro next time. |
| Groups used too much cheesecloth in activity 2 and we ran out. (Extra cheesecloth in the "Lab 2" drawer in the stockroom.) Show them the right amount to use (about the size of a playing card). |

| |
|---|
| What questions did students ask? |
| Why do the starch and the sucrose solutions have different results? I had them look up the structures on their phones and then they were able to answer it themselves. |

| |
|---|
| Additional Notes: |
| Schedule: Quiz (10 min), lecture intro (10 min), groups work on activities 1, 2, 3 (80 min), re-group and share data and graphing (20 min), start seeds for genetics lab (15 min). |
| I was more relaxed than week 1! I got to know the students a bit while they were working, and they seemed more willing to ask and answer questions. |

CHAPTER 4

# Content: Be a Guide for Students

I F YOU KNOW EVERYTHING THERE is to know about your discipline with complete certainty, skip this chapter. If you are a real person, keep reading.

You will not be an expert in everything you teach. In a semester-long science course, you would be lucky to have deep content knowledge for just a few of the topics. This is especially true for an introductory-level course that covers a broad subject. At my institution, plant and animal biology are combined in one introductory course. The TAs for this course are either botanists or zoologists, so *every* TA starts out feeling that at least half the course is somewhat unfamiliar to them.

And that is OK. In fact, a TA's lack of expertise can be *beneficial* to students. (More on this later.)

A common concern for new TAs is how to prepare to teach unfamiliar content. ("Content" refers to the ideas, facts, and techniques that you will teach, and the students will learn.) Even if you are teaching a course in which you have some level of mastery, there will be areas where you are less confident. Every new teacher finds themselves in the position of having to learn or relearn a topic before turning around and teaching it. It is helpful to have a system to help you prepare to teach new content. This chapter will give you that system.

It will take some time—learning something new always does. However, preparing to teach something new can, ultimately, be satisfying. Most TAs in the plant and animal course described above quickly become good at teaching the whole course. They enjoy teaching those topics that were initially unfamiliar.

## BE A GUIDE, NOT AN EXPERT

Not being an expert in the content you are teaching is *not* a bad thing. In fact, it can be positive. Having to learn or relearn some of what you are teaching will make you a better teacher. As you learn the material in preparation for teaching, you will identify what is most challenging and develop strategies to learn the hard stuff. Those insights will help you help the students. The students might learn a topic *better* if you are learning along with them.

However, it is not always easy for a new TA to adopt this mindset. Some new TAs deal with imposter feelings. Fear of being exposed as an imposter is heightened when you do not feel like you know the content well. A new TA might look at upcoming content in the class and think, "I barely remember learning this myself, so how can I be qualified to teach it?" Such thoughts can intensify imposter feelings.

To break a cycle that amplifies imposter feelings and erodes your confidence, remember that the students need you to be a *guide*, not an *expert* (see Box 1). Students must learn the ideas, facts, and techniques for themselves. Even if you are the undisputed worldwide expert on a particular subject, you could be an ineffective teacher of that subject if you can't be a guide for student learning. Being a good guide for students is less about content and more about helping students stay organized, giving them strategies for studying, and providing encouragement. (It is, of course, possible for someone to be an expert *and* be an effective guide for student learning. Educational institutions are full of such individuals—experts in their fields who are also thoughtful and dedicated teachers. Hopefully, you will encounter many such people throughout your life.)

Your goal is not to become an expert in the days before you teach a new topic. That's unrealistic. Rather, your goal is to learn the material and then help students learn it. The fact that you have recently been through the learning process will make you a more effective guide. As you prepare to teach new content, especially if it is something you are not very familiar with, remind yourself of the following: you know the fundamental principles of your discipline, you have the academic skills to learn new material, you don't need deep knowledge of the topic to teach it at an introductory level, and students need you to be a guide—not an expert.

### BOX 1. BEING A GUIDE FOR STUDENT LEARNING

The best teachers facilitate student learning. They put students in a position to be successful by breaking down a process into manageable steps, by ensuring that students have the necessary preparation, and by providing help when needed. Remember, you cannot learn it for them!

| A POOR GUIDE... | AN EFFECTIVE GUIDE... |
|---|---|
| Makes authoritative statements. ("The common features of eukaryotic cells are a cytoplasm, nucleus, and plasma membrane.") | Makes suggestions and asks leading questions. ("The table comparing plant and animal cells is very useful. Can you list the features both cells share?") |
| Only gives students the "take-home" message. | Provides information to help students reach the conclusion themselves and helps summarize. |
| Assumes the audience has the background knowledge to understand the core concept. ("As you know, there is an inverse relationship between pressure and volume.") | Explores the audience's level of knowledge and helps to fill gaps. ("Do you remember the ideal gas law from chemistry class? Can we write it out? What does it tell us about the relationship between pressure and volume?") |

## JUMP IN THE (NOT-VERY-DEEP) END

There are no shortcuts to learning or relearning content in preparation for teaching. The preparation can be manageable if you follow these steps:

1. Identify the minimum needed to help you feel confident as a teacher.
2. Study with the goal of building confidence in this minimum material.
3. As you study, stay alert for insights that will help students.
4. Build knowledge over time as you reteach the material.

This may sound a little like an endorsement of a "fake-it-until-you-make-it" mentality. However, you aren't trying to fool anyone into thinking you are an expert. You are building your own knowledge with the goal of helping students build their knowledge. Let us break down each of the steps listed above.

*Identify the minimum:* There are probably between one and three key ideas the students need to learn about the topic at hand. These may already be written out in the textbook or lab manual. If they aren't already written out, consider the level of understanding expected for quizzes, exams, and written reports, then work backward from there. Trust your intuition and experience—it may have been a while since you were first exposed to this content but your recollections of what was important *then* can help you *now*.

***Build confidence in the minimum material:*** Study with the goal of being able to clearly explain the minimum information you identified in the first step. Avoid doing a deep dive into the material—there will be time for that in the future. **Do not memorize!** You will access your notes while you are teaching. If you find yourself worrying, "How will I keep all this information straight?" plan to write it on the board or project it on the screen during class. For example, in lab, you are using an equation with lots of variables. Rather than take the time and energy to memorize what each variable represents, write each variable and a short explanation on an unused end of the chalkboard, or put them at the bottom of a slide you are projecting. Your students will think this was done for their benefit—and they'd be right! You may have been motivated to do this as a confidence builder for yourself, but for your students, it's a helpful resource and a model of how to stay organized when doing calculations.

***Stay alert for insights that will help students:*** Because you are doing this work with the goal of teaching, your "teacher brain" will inevitably be switched on during the process. Let your teacher-self watch over the shoulder of your student-self. Be aware of what academic skills you are using. Are you skimming the textbook looking for key words? Are you visiting trusted websites for additional information? Insights into your learning process can be shared with students when your class meets.

***Build knowledge over time:*** Your preparation has given you a good foundation to build on. Next term, if you are teaching the same class again, consider how you might add to your knowledge. Would it be helpful to find additional examples to add interest to the material or make it relevant to students' lives? Was there a part you found interesting and that you wish you knew more about? You can always supplement the minimum. Deeper content knowledge will enhance your teaching and strengthen your confidence.

## SHARE YOUR PROCESS, NOT YOUR PRODUCT

Imagine a cooking show in which a celebrity chef describes a fabulous dish and then walks to a table where the finished dish sits hot and beautifully presented. The chef's expertise might be unquestionable, and the food might be delicious, but the show would be of no value to anyone. Value comes from seeing the process—how thick to slice the onions, how to tell when the pan is hot enough, etc.—not just seeing the final product.

New teachers can be reluctant to admit that they are not experts. They worry that showing any gaps in their knowledge will result in losing the respect of the students. They spend much more time preparing than is necessary because they are memorizing terms, equations, and processes. In the classroom, they share only the *product* of their learning with the students—they just share the finished meal and not the steps

involved in its preparation. In a classroom, the result of only sharing the product is a tightly controlled lecture in which the teacher does all the talking, questions are discouraged, and the students are less engaged. Ironically, the person who learns the most in this scenario is the teacher!

There is a better way—one that lessens the pressure on the TA and helps students develop as learners. Consider sharing your learning *process*. You have arrived at your current position because you have proven to yourself and others that you are capable of learning hard things. As you prepared for teaching, what did you struggle with? How did you work through those struggles? How did you stay organized? What sections of the lab manual or textbook were especially helpful to you? How did you set up problems before solving them? The students benefit if you choose to share this information. The benefit comes from seeing the way an experienced student tackles a hard or new topic. It is OK to say, "For this topic, I'm learning right along with you." Here are other things you can say as you share your learning process with students:

- "I'm not an expert in this topic and I learned a lot in preparing for this lab/discussion."
- "One thing I found very helpful, was (making a table, looking at a figure in the textbook, diagraming the steps, using the textbook glossary)."
- "I found this (figure, resources, etc.) to be helpful and maybe you will too."
- "While I was learning more about this topic, I found I often referred to this equation. I'll write it on the board so we can all access it."
- "Good questions! I found that part hard to understand too. After thinking about it for a while I realized that…"
- "That's a great question and something I haven't considered. I'll investigate it and share what I learn with you later this week." (Follow up with an e-mail to the class.)
- "Good question. I think I'll need to consult an expert on that and report back to you." (Make sure to follow through!)
- "I don't know, but I'll find out."

There are many benefits to sharing your learning process. For you, it can remove the pressure of having to appear as an expert. This approach also builds trust between you and the students in your class. The benefit to students is that in addition to helping them learn the topic at hand, they are also seeing your academic skills at work. Many students may not yet have learned how to study for college-level classes, or they do not know how to adapt their studying for a new discipline. They might not know how to use all the resources in textbooks (tables, figures, index, glossary, etc.) to find helpful information. They may not have developed the skill of thinking critically about how well they understand what they are reading. By sharing your learning process, you are modeling a growth mindset for students. They benefit from seeing someone in your position acknowledge that a topic is difficult and then work hard to understand it. "This is complicated, and I had to struggle with it. Here's how I did that... ."

## CASE STUDY: THE FRUIT LAB

Marco is a graduate student studying small mammal ecology. He is a laboratory TA for *Introduction to Anatomy and Physiology*. About half of the labs for this class cover content from plant biology.

This week's lab will focus on fruits, and Marco is anxious. He took a similar class as an undergraduate many years ago. He remembers looking at plant samples but does not recall any of the terms or concepts. The lab manual and textbook have figures and tables listing about 30 unfamiliar anatomical terms (mesocarp? drupe?).

At the weekly TA orientation meeting, Marco learns what plant samples will be on display for the students to examine. More importantly, he gains insight into the main learning objective for the lab—what the students are expected to understand and be able to do. The focus is on recognizing the three main fruit anatomical categories (simple, multiple, and aggregate). This is confirmed by a look at the lab practical exam. Marco sees that students will be tested on their ability to correctly sort unfamiliar fruits into these categories. Marco has **identified the minimum** needed to be an effective guide for his students. He needs to know the anatomical categories and he needs a good strategy for distinguishing between them. In addition, he needs to be able to apply this strategy to the fruits that will be on display in the lab.

Marco begins to **build confidence in the minimum material** as he prepares to teach. At the same time, he will **stay alert for insights** he can use to help the students. Skimming the textbook, Marco finds the pages and figures that provide the information he needs, and he jots the page and figure numbers down. Marco knows from his experience in animal anatomy labs that a table is helpful when one needs to compare and contrast, so he draws a large table on a sheet of paper that will become part of his Personal Teaching Document. In the back of his mind, Marco thinks he will probably draw a similar table on the whiteboard in the lab to help the students stay organized. The table has one column for each of the fruit categories. At the top of each column, he writes a shorthand definition of each category based on the definitions in the textbook. If he does end up writing this on the board, having short, accurate definitions in his teaching notes will be key. As he writes the definitions, he realizes that all are dependent on another term, "ovary." He returns to the textbook for a precise definition of that term and writes that under the table. At this point, his teacher's brain clicks on. He imagines starting off this activity saying, "While preparing for this lab, I realized that all the fruit categories were dependent on understanding what a plant ovary is, so let's define that." In his teaching notes, he draws a star next to the ovary definition and writes "DO THIS FIRST!" in the margin.

Now Marco applies his new knowledge to the fruits that the students will examine in lab. As he categorizes them, Marco writes the names on the table in his notes. Obviously, he will not write these answers on the board in lab—the students need to figure it out for themselves. However, he knows that having this information in his table as a reference is important for his confidence. As he categorizes the lab fruits, Marco practices in his head how he will engage students during the activity: "What category did you put the cherry in? That's right, what did you notice about the cherry that helped you identify it as a simple fruit? Can you rephrase that but use 'ovary' in your answer?"

Marco is *not* memorizing. During his introduction, he will write important terms on the board for everyone, including himself, to see. Teaching from memory would be a waste of time and energy.

He will access his teaching notes as he interacts with students. After this preparation, he will be familiar with all the information and prepared to be an effective guide.

This process has not taken much time. Marco simultaneously learned the material for himself and developed some good ideas to use in the classroom. Is Marco a fruit anatomy expert? No. Will he be able to answer every question that comes up in lab? No. But, Marco is prepared to guide his students through this activity.

Marco has used his own academic skills to reach this point (scanning the textbook, organizing information in a table, and self-testing). He plans to **share his process, not his product**. He will keep his product—a list of fruits sorted into categories—to himself as a reference. Marco's teaching plan involves sharing the strategies that he used while preparing, and he can be open about the process. For example, before he draws the blank table on the whiteboard, he'll say, "I've always found that making a table is really helpful for tasks like this."

In addition to gaining confidence to teach the Fruit Lab, Marco may have become more interested in the topic. After all, it's hard not to become at least somewhat engaged with a topic you are studying. Marco may find it satisfying to **build his knowledge over time**. The next time he teaches the fruit lab, he could seek out interesting examples or rare exceptions to the basic categories. Maybe researching the answer to a student's question will enhance his teaching materials. Regardless of the motivation, Marco has a good foundation for teaching the lab to which he can add useful information in the future.

## A SYSTEM FOR PREPARING NEW CONTENT

This section gives you a system for learning or reviewing content in preparation for teaching. (Good news—it is easier than preparing for a test!) The best time to do this work is after you have had some exposure to the lab. Wait until after the weekly coordination meeting or after you have observed another TA teaching. If these resources are not available to you, skim the lab manual, class texts, or whatever materials you have. If you try to prepare without this glimpse into what the week's session entails, you will not spend your time wisely. This is not unlike studying for a quiz or test. If you study an entire chapter in detail before attending the lecture, you will probably find that at least some of the material you covered was unnecessary.

What follows is a series of questions to ask yourself as you prepare to teach new content. These questions incorporate the guidelines outlined earlier in this chapter. The questions are presented in the form of a worksheet. (Download the worksheet at TheTAsGuide.com.) Use the worksheet if it's helpful to you. Even if you do not fill in all the boxes, these questions can still guide your preparation. In addition to a blank worksheet, a sample for a biology topic is included. (Note: Consider breaking the class content up into more manageable "chunks" of content and go through this process for each.)

## CONTENT PREPARATION GUIDE

| |
|---|
| Topic: |
| What helped *me* learn this (before as a student, now as an instructor)? What from my learning process can be shared with students? |
| What do the students need to learn? At what level (introductory, advanced, etc.)? |
| Why do they need to learn it? |
| How will they learn it? |
| How can I help them learn it? |
| If possible, list one or more interesting facts to share. |
| Notes and reflections (before, after, and during class). |

## EXAMPLE CONTENT PREPARATION GUIDE

| |
|---|
| **Topic:** Levels of Protein Structure |
| **What helped *ME* learn this (before as a student, now as an instructor)? What from my learning process can be shared with students?** <br> Reading textbooks, having the basic amino acid structure memorized, making a table comparing each of the levels (primary, secondary, tertiary, quaternary). |
| **What do the students need to learn? At what level (introductory, advanced, etc.)?** <br> Introductory level: What bonds are associated with each level (covalent, ionic, hydrogen). Protein function is dependent on protein structure (conformation). |
| **Why do they need to learn it?** <br> Lecture exams and homework. <br> For Enzyme Lab—explain the effect of heat and pH on the structure (denaturation). |
| **How will they learn it?** <br> Lecture, textbook, pre-lab lecture. |
| **How can I help them learn it?** <br> Show how I set up my table (row and column headings). Students fill in the table working in groups. Draw attention to the importance of upcoming labs. When possible, ask "What level of structure are we focusing on?" |
| **If possible, list one or more interesting facts to share.** <br> A single amino acid change (primary structure) in hemoglobin leads to sickle-cell anemia. The secondary and tertiary structures of hemoglobin are affected leading to chains of protein forming. These warp the red blood cells into a half-moon (sickle) shape. |
| **Notes and reflections (before, after, and during class).** |

## ONE WEEK IS ENOUGH

New teachers often put pressure on themselves by thinking that they need to have the entire course planned from day one. A TA teaching a laboratory section for the first time might feel unprepared if they do not have a plan for every week, all their quizzes written, and the lab exam created.[1]

Not only is this level of preparation *unrealistic* for a new teacher, but it is also counterproductive. It is not a good use of your time to write a detailed teaching plan for the entire course until you have met the students and started doing some teaching. You will learn a lot about the students during the first week or two of the class. You will learn what they know. You will discover how much you can expect them to learn independently and how best to guide them. Your teaching plans will evolve the more you know about the students.

You will also learn about yourself during the first weeks. You will learn how much preparation time you require before teaching. You will learn the best way to organize your teaching materials. Your preparation will change—and become more efficient—with this knowledge.

**WHAT ARE YOUR INSIGHTS ABOUT PREPARING NEW CONTENT?**

"I WAS VERY ANXIOUS. WATCHING SOMEONE ELSE TEACH—THAT WAS WHAT REALLY HELPED ME TO CALM MY NERVES BY SEEING WHAT I WAS GOING TO BE DOING AND WHAT IT MIGHT BE LIKE."
  -Katie, 2nd year graduate student

"YOU'RE NOT EXPECTED TO LEARN OR KNOW EVERYTHING, BUT YOU'RE EXPECTED TO BE ABLE TO FIGURE IT OUT. I REALLY THOUGHT THAT I NEEDED TO BE AN EXPERT IN EVERYTHING THE LAB WAS SUPPOSED TO BE ABOUT. BUT I DIDN'T. I JUST NEEDED TO BE ABLE TO GUIDE THEM THROUGH THE ACTIVITY. IF THEY HAVE QUESTIONS THAT ARE BEYOND ME, I JUST NEED TO HAVE A RESOURCE-KNOW THOSE RESOURCES TO ANSWER THAT QUESTION."
  – Annie, 2nd year graduate student

"I THOUGHT IT WAS A WEAKNESS SAYING, 'HOLD ON, I WILL GO FIGURE IT OUT.' BUT IT TURNS OUT, IN MY [STUDENT EVALUATIONS], THEY APPRECIATE THAT I TAKE THE TIME TO SAY I DON'T ACTUALLY KNOW."
  – Kerry, 2nd year graduate student

---

[1] The advice in this section assumes that you are teaching a class with an established course schedule and syllabus in place. A great deal of more work is needed if you are responsible for planning the class and writing the syllabus. This level of curriculum planning is beyond the scope of this book, and you are encouraged to seek the help of other teachers for advice and sample syllabi you can use as a model.

If you do lots of planning without the benefit of the knowledge you gain in the first weeks, you will likely find that your plans will need extensive revision to be useful. Another advantage of preparing and planning week-to-week is that you will have the opportunity to build on student experiences. At the outset of the semester, it is hard to envision how themes and skills will build over time. For example, while preparing for class in Week 5, you might realize that the problem-solving students will do is an extension of the type of problem solving introduced in Weeks 2 and 3. You can take advantage of this insight to make your preparation more efficient and to help your students make connections across the course.

Your goal, the first time you are teaching a class, should be to stay about a week or two ahead of the class. Even that may not be possible and just staying a day or two ahead of the class is fine. Not only is it okay to plan week-to-week, but it will probably be the most efficient way to prepare.

It will be helpful for you to know the general class schedule and to have an overview of the course in mind. Know what topics are covered and in what order. Get a sense of how the material builds from one week to the next. But do the detailed planning as you go.

CHAPTER 5

# Inclusivity: Help Students Feel Welcome

ATTITUDES ABOUT EQUITY IN HIGHER education are evolving. Previously, the idea of a level playing field was accepted: "The students all hear the same lectures, do the same labs, and take the same tests. My door is always open to anyone who needs help."

Instructors were aware that students entered college with different levels of preparation. But many instructors operated on the idea that anyone who worked hard enough and was clever enough could succeed. Today, this attitude feels naïve, and it is urgent for institutions and teachers to address inequity in higher education.

Author and researcher Bryan Brown (2019) compares educational inequities, especially those experienced by black and brown students, to an unfair tax—an extra burden applied to some students and not to others. A student may struggle in college classes not because they are deficient, but because they are taxed. The extra burden may be a feeling that they don't belong or that they aren't capable. The extra burden may be a worldview that equates asking for help with personal weakness. The extra burden may be from hunger or working two jobs or childcare. It is true that all students hear the same lectures and take the same tests. But can we expect equal success when some students carry extra burdens into the classroom?

Failure to acknowledge and address inequity has led many promising students to leave academia. This is especially true in STEM (science, technology, engineering, and math) fields, effectively silencing voices that could have made great contributions to science and engineering. One of the most important factors affecting whether students, especially students of color, persist and earn a degree is the quality of their interactions with instructors (Carter and Wilson 1994). Through their interactions with students, college teachers can positively or negatively affect a student's trajectory. "Because educators have the potential to set the tone for their students' experience," writes Professor Dena Samuels, "it is important that we take account of the [educator's] attitudes and behaviors" (Samuels 2014). As teachers, we must try to make all our interactions count and do our best to avoid interactions that send discouraging or exclusionary messages to students.

As a teacher, you are a role model. If you are from a privileged group, you can model antiracist language and actions. If you are a member of a group that has been traditionally underserved by the college system, you can empathize with and support students from similar backgrounds to your own. We all have a role to play in supporting students as they pursue their personal, academic, and professional goals.

"SOMETIMES I HAVE TROUBLE PICTURING MYSELF IN THIS FIELD."

Educational institutions have a long history of excluding people based on race, ethnicity, gender, and other factors. This chapter does not offer a solution to systemic racism or sexism in education.[1] Hopefully, it will raise your awareness of how inequities show up in the classroom. The chapter suggests actions you can use in your teaching that can help all students feel like they belong and can be successful. Samuels reminds us, "The classroom can be a laboratory for learning how to engage and connect with one another across our social differences." When we identify barriers to inclusion and do our best to lower those barriers, we become more conscious of the diversity around us as teachers and as people.

## BEING AN INCLUSIVE TEACHER

It is important to define "inclusion" and describe what it means to be an "inclusive educator." To arrive at a good description of an inclusive teacher, we must define some other ideas that are closely aligned with inclusion: diversity, intentionality, and equity.

---

[1] If you are interested in promoting institutional change, you are likely to find many allies at your school who share this goal. Ask if your department or college has committees on inclusivity or social justice. Your university likely has a center devoted to teaching and learning that may offer workshops or book clubs focused on inclusive teaching in the college classroom. These are good ways to connect with other teachers committed to promoting social justice at the institution.

*Diversity:* Discussions of diversity on college campuses initially focused on race and gender. To be sure, race and gender are important factors that contribute to diversity. The history of race-based and gender-based exclusion at academic institutions makes these factors deserving of that attention. But diversity can and should be more broadly defined. A typical college class will include students that differ in a number of characteristics including race, gender, age, sexual orientation, socioeconomic status, learning style, physical health status, mental health status, religion, degree of introversion, and more. All these characteristics contribute to the diversity of the individuals who will walk into your classroom.

*Intentionality:* It is easy to fall into teaching styles that don't respond to student diversity. "Default" teaching styles tend to put the teacher at the center of the action. Teacher-centered instruction sends a message that the teacher and the content are most important in the course. Why does a teacher-centered style tend to be the default in many college classes? First, it is what many of us experienced as students, and many associate this style with "good" college teaching. Second, the layout of most college classes promotes a teacher-centered style with the teacher "onstage." But the main reason a teacher-centered style is common is that teaching this way is easier to prepare and less risky to do. The teacher simply follows their predetermined "script" and student questions are rare. (In Chapters 6 and 7, we explore ways of teaching that make the classroom more student-centered.) Inclusive teaching starts with the teacher deciding to do things differently. The question is not "How can I cover the most material possible?" Rather, the inclusive teacher may ask, "Is the way I'm teaching working for the students? Are there changes I can make that help more of the class be successful?" These questions lead to intentional choices about how best to teach.

*Equity:* In an equitable classroom, each learner has access to the resources they need to succeed. Creating an equitable classroom requires the teacher to recognize the different characteristics of the students and do their best to create learning opportunities for all.

So, what does an inclusive teacher do? An inclusive teacher recognizes *diversity* in their classes and takes *intentional* actions to make the

"YOU'RE MAKING IT HARDER THAN IT NEEDS TO BE."

educational experience *equitable* for all students. The inclusive teacher works to make sure every student feels as though they belong in the class and that they can succeed.

## HAVE AN INCLUSIVE MINDSET

The most common teaching assignment for new TAs is as part of a group of instructors helping to deliver a large-enrollment course. You know these courses: Bio 101, Chem 123, Physics 1A, Intro Art History, etc. These courses typically feature a large lecture with hundreds of students supported by smaller laboratory or discussion sections led by TAs. Large-enrollment courses may be the *least* inclusive classroom environment a college student will experience. (Paradoxically, this environment often serves as the entry point for new students such as freshman and recent transfer students.)

It is very difficult to create a course structure that serves a large number of students while still valuing and supporting all individuals who make up the class. Large-enrollment classes tend to have rigid policies relating to absences and late work. Flexibility isn't an option. Assessment in large-enrollment courses focuses on multiple-choice testing as opposed to more diverse forms of assessment. Instructors and TAs for large-enrollment courses have fewer opportunities to modify content and instruction based on student strengths and weaknesses. All these factors can make students in large-enrollment courses feel "like a number" or that they must "sink or swim." The format disadvantages students who do not easily make social connections in class and fosters imposter feelings (more on this later).

Instructors may lament the fact that large-enrollment classes do not have the level of inclusivity that they themselves value. TAs involved in these classes can feel a frustrating tension between upholding the course policies, which they typically have little control over, and their own desire to help students feel welcome and successful. Can a TA who values inclusivity have much of an effect given these limitations?

Yes! The lab or discussion sections associated with a large-enrollment class can be the place where students have the greatest chance to feel seen and heard, and where they feel more connected to the class and their peers. As a TA responsible for some of these sections, you can play a role in making a large course feel more personal and welcoming. The most important thing a TA can bring to the class is an inclusive mindset. A TA with an inclusive mindset will (1) value every learner in their class, (2) learn who the students are, (3) share their experiences in ways that help students navigate the class, the major, and college in general, and (4) talk about inclusion—and not just on the first day. Even if the course treats students "like a number," you can treat the students in your section like the diverse individuals they are.

## TAKE ACTION

What follows is a list of actions you can take that make a classroom more welcoming and inclusive. Everything listed below is an *intentional* action—the TA is making a choice to put students first. The actions are roughly organized by the amount of effort they require. Actions near the top of the list require almost no additional effort. Further down the list, you will find actions that require more thought and preparation but that deliver greater positive impacts for the students. If you are a first-time teacher, focus on actions that require less effort. These *will* make a difference! As you gain experience, you will find it easier to incorporate more of the actions on this list.

Is this list of actions complete? Of course not. There are many things an instructor can do to make the classroom environment more welcoming, and only a small sample is listed here. Perhaps reading this list will prompt you to come up with new ideas to help students feel welcome and included. (The resources listed at the end of this chapter provide more suggestions for fostering inclusivity.)

Is it essential to take every action on this list? Of course not. Some will come naturally to you and can be easily integrated into your teaching. Some actions will require considerable thought and preparation. By taking any action on this list, you make the classroom more welcoming and inclusive.

- **Welcome students** as they arrive at class. Smile and greet students rather than retreat to behind the instructor's desk or podium. Being relaxed and available at the start of class is easier if you feel well-prepared. This might not be possible to do every week of your first term as a teacher. Do your best to find time to acknowledge each student as they arrive.
- **Talk about inclusion and diversity.** Repeat this message throughout the term but especially during the first few weeks. You can say things like: "Everyone should have an equal opportunity to be successful. Please speak with me if anything is limiting your opportunity to succeed," or "None of us are starting from the same point. But everyone can be an asset in class, and everyone can improve," and "Everyone will feel uncertain or confused at some point. Give each other support and plenty of chances to ask, talk, and explain."
- **Learn student names** and how to pronounce them. There are many reasons to learn student names. One important reason is that using a person's name shows respect and builds human connections. In a classroom setting, doing your best to learn names in the first few weeks of the class shows the students that you see them as individuals. (A preclass survey like the one described later in this chapter can be a great tool for asking students to help you with pronunciations.)
- **Share your personal pronouns** and encourage students to share their pronouns. Provide multiple ways for them to share their pronouns with you if they choose to do so (e.g., individually, via a survey, paper name tents, etc.). There is more about pronouns later in this chapter (see Box 1. Pronoun Pressure).

- **Ask students how they are doing** and then *really* listen to the answers. Consider asking, "How is your week going?" or "What's on your mind this week?" Ask "How did you get to class today?" or "How was it working on the latest assignment?" Asking someone how they are doing and really listening to the response is a high form of respect.

- **Share your experience** especially if it relates to the class or to college in general. By sharing an experience where you faced a challenge in college, you normalize these experiences and identify yourself as a resource for students who may be facing similar challenges. What strategies did you use to study for the sort of content you are teaching? If you were a commuter student, how were you able to feel more connected to campus culture? How did you use office hours to improve your experience in classes? You have a wealth of knowledge that can help students figure out college. Sharing stories about your own life and learning builds trust between you and the students. Work these stories in as you teach.

- **Acknowledge current events** that may be affecting student well-being. It can be tempting to think of the classroom as walled off from the outside world—a space where we can just think about molecules or music or circuits in isolation. But it is unrealistic to expect you or the students to simply tune out news of tragedy, political unrest, or acts of hate. Acknowledge these events but don't put pressure on yourself to relate them to the course content or "put it all in perspective." It is enough to just provide time for thought. For example, you can say, "I was saddened by this news, and I imagine many of you are too. Before we begin, I'd like everyone to take one minute to reflect silently on recent events and how they are affecting you and members of our community." You can also offer to be available after class for anyone who wants to talk.

- **Talk about imposter feelings.** (More on this later in this chapter.)

- **Use inclusive language** when teaching. (More on this later in this chapter.)

- **Give everyone a chance to participate.** Do not call on the same students over and over. Say "I'd like to hear from someone who hasn't said anything yet today."

- **Reduce barriers to participation** by incorporating any of the strategies in Chapter 6. The Think–Pair–Share strategy and its alternatives described in Chapter 7 can also be a way to help more students feel comfortable sharing their ideas.

- **Be flexible when you can.** For a class that has strict rules about when assignments are due ("Turn in at the start of class or not at all!") consider reasonable alternatives. For example, could you accept a screenshot or cell phone picture of the assignment when a student is delayed or otherwise unable to participate? (Small acts of flexibility like this are usually fine and at the discretion of the TA. However, if you are part of a team of TAs teaching a large-enrollment course, check before offering greater flexibility, such as alternative testing dates, or opportunities to turn in late work at the end of the term.)

**Explain teaching decisions** intended to increase inclusivity. "For this activity, I'm going to randomize you into groups of four or five. I'm doing this because I think it's important for you to get to know students in the class you might not sit next to. Please introduce yourselves before you start working on the activity."

**Deemphasize competition** and encourage collaboration. Remind your students, "You aren't in competition with each other. In fact, having a positive attitude about working with your peers will help you be successful in this class."

**Reach out** to absent students or students who are struggling with the material. This can be done in the form of an e-mail: "You weren't in class this week. Is everything OK? I'm eager to help you get caught up. My office hours are.... ." or "It looks like the last quiz was difficult for you. I know you can do better and I'm eager to help you get there. Is there a way I can help?"

**Help students make social connections** within the class. Any student can feel isolated. Transfer students and students who commute to campus are especially vulnerable to isolation (Townsend and Wilson 2009). Helping these students make connections within the class can improve their college experience academically and socially. Helping can be as simple as introducing two students who are both looking for someone to study with. Introduce students to each other when they are in office hours at the same time. On the back of a paper quiz, you can include "I'm eager to help you make academic connections in the class. By writing your e-mail below, you grant me permission to share your e-mail with others looking for study partners."

**Hold some office hours in the evening** or late afternoon maybe via videoconferencing. If your schedule allows, consider holding some of your regular, required office hours (not *extra* office hours) in the evening. This timing could make it easier for students with long commutes or difficult work schedules to participate.

**Make a list of campus resources** that help students deal with emotional, social, or financial issues and share that list with the students. Say "I want you to stay happy and healthy this term. If you find yourself having a tough time, let me put you in touch with the campus services that can help." A template is provided for you to collect this information in the Appendix: Guidelines for Teaching.

**Send a welcome e-mail or video** prior to the start of the term. Making students feel welcome and included can begin before the students come to the first class. A welcome e-mail or video can give you the opportunity to:

*Introduce yourself:* "I'm a grad student studying [topic]. In my spare time, I like to [your hobby here]."

*Share your excitement:* "I'm excited for you because in this course you will learn interesting and important things about [topic]."

*Address student anxiety:* "You might be feeling a little nervous because there is a lot to learn. Remember that no one is born knowing [topic] and we will all learn together. I'll give you lots of guidance and opportunities to practice."

*Let the students know you are excited to meet them:* A video can be a great way to do this. Keep it simple! Use a phone or a laptop camera and distribute it through your course website, e-mail, or YouTube. Write an outline, film a maximum of two takes, then use the best one. It doesn't have to be perfect.

**Include a diversity statement** in the course syllabus if you have the option to do so. For many large-enrollment courses, there may be a standard syllabus that all TAs use. If that syllabus has a diversity statement, point it out to the students on the first day so they see that respecting diversity is as important as the exam schedule, grading scheme, and all the other parts of the course you cover on day one. If the syllabus does not have a diversity statement, add one (if you are permitted to modify the syllabus) or share it orally or as a slide. Here is a sample:

"Our class should be a supportive community in which we work together to learn about [topic]. Diversity of background, identity, and perspective strengthens this learning community. All class members should contribute to a respectful, welcoming, and inclusive environment in the classroom and support the efforts of your peers to learn."

**Do a preterm survey** to learn about your students, their attitudes toward the course, and the challenges they anticipate. A sample preterm survey is presented later in this chapter.

**Complete SafeZone Training** to learn how to best support LGBTQ+ students in your classroom. If in-person SafeZone training is not available at your institution, the SafeZone Project offers an online certification course. (See resources at the end of this chapter.)

**Advocate for changes.** If you are part of a team of TAs teaching sections of a large-enrollment class, you can advocate for changes in class policy that increase equity. Because you work closely with students, you might be aware of problems that the supervising faculty member does not see. Maybe the class policy is to give a 10-minute quiz beginning exactly at the scheduled start time for each lab section and students arriving late have less time to work. You observe that students who take the bus are being unfairly penalized because of bus schedules beyond their control. Bringing up these barriers to student success can lead to improved inclusivity across the whole class.

There is much to consider when trying to make teaching more inclusive. The list you just finished reading gives you some ideas of in-class actions you can take. In the following sections, we will take a closer look at a few of these steps toward greater inclusivity.

## HELP STUDENTS WITH IMPOSTER FEELINGS

The fact that you are in your current position (graduate student, lecturer, advanced undergraduate, etc.) means that your experience on college campuses has been largely positive. You have felt included and successful. Or, if you *haven't* felt included and successful, those feelings have not stopped you from achieving success—and you will be an amazing role model for students. For those of us for whom college was a positive environment, we must be conscious that the students we teach will not necessarily experience college in the same way. They may feel disconnected or unwelcome. Like college "isn't for them." They may feel like they aren't "college material."

Imposter feelings[2] can affect any student at any time regardless of race or ethnicity, gender, or economic background. Feeling like an imposter—like you don't belong in college, like you've managed to fake it so far, like your success isn't earned or deserved, like you will inevitably be exposed as a fraud—is very common among college students. Multiple studies agree that most students experience imposter feelings during their time in college. No one is immune. However, feeling like an imposter is more common among students who identify with groups that have been traditionally underserved by higher education (Canning et al. 2020). These groups include black and brown students and women in STEM disciplines and nontraditional students such as older students or career changers. Awareness that imposter feelings might affect some students disproportionately is a step toward inclusive teaching.

Knowing that many of your students will experience imposter feelings, how can you help? There is no quick fix; however, psychologists have identified some fairly simple ways to reduce imposter feelings (adapted from Orozco et al. 2023 and Palmer 2021).

***Raise awareness of imposter feelings:*** Students in your class may not be aware that what they are feeling is common. At the core of imposter feelings is the belief that just

---

[2] The more common term, imposter syndrome, is avoided here. The word "syndrome" has a strong medical connotation sending a message that feeling like an imposter is an illness. Many educators use "imposter feelings" or "imposter phenomena" to avoid further stigmatizing students who sometimes feel like they do not belong.

you are feeling this way and that everyone else feels happy and successful. Remind the class that imposter feelings are common, that you understand, and that you are willing to help. Say to the class, "Many students struggle with the feeling that they don't really deserve to be here or that they will eventually be 'exposed' as a fraud. These are imposter feelings and they're really common. You are not an imposter, and you deserve to be here. I'd be happy to talk with any of you about these feelings." Have you experienced imposter feelings as a learner or as a teacher? Sharing your story, if you are comfortable doing so, can go a long way toward quieting imposter feelings in your students.

*Encourage collaboration:* Imposter feelings thrive in isolation. As an instructor, you can encourage students to collaborate during class time and study together outside of class. Collaborative learning reduces the perception that students are in competition with one another. Feeling that you are in competition with those around you can amplify imposter feelings. The effect of perceived competition is stronger in among students who are the first in their families to attend college (Canning et al. 2020).

*Help students recognize how much they have learned:* If a student takes you up on your offer to talk about imposter feelings, you can help them recognize how much they have learned and achieved. Remind them that no one is born knowing genetics, organic chemistry, physics, or whatever the subject of your class is. Next, remind them of their growth so far. Make a list of topics or techniques they've already learned in the class. Remind them, "You are learning a lot of hard things, and it is normal to feel confused sometimes. Learning hard things is what you came here to do. And you are doing it."

Students may be especially likely to seek you out after a poor academic performance (a low quiz grade, a failed exam, etc.). Ensure them that a poor grade on an assignment or test does not define them. Their poor grade most likely reflects inadequate preparation—underestimating the time needed, starting too late, or simply not practicing the right things—and not a lack of ability or caring. Try to help them move forward in a positive way by identifying what went right and then helping them develop strategies to improve in the areas that were challenging for them.

Of course, imposter feelings are not confined to undergraduates. You may also have imposter feelings related to starting graduate school, finding your place in a research group, or being a first-time teacher. The strategies described above can also help you. Know that you aren't alone in feeling this way. Build a network of supportive people. Try to move on in a positive way from setbacks because these do not define you or your ability. Do not hide from validation—give yourself a pat on the back when you deserve it. You are doing a good enough job!

## WORDS MATTER

The words you use when teaching influence whether students feel included or excluded. As you read this, you are probably thinking, "I'm not going to ridicule students! I won't say things in my class that are demeaning or inflammatory (racist, sexist, etc.)!" Of course you won't. However, even small, casual comments to students, regardless of how they are intended, can send a message that some students are more welcome in class than others. These small statements, called "microinequities," are usually unintentional. Microinequities have a cumulative effect. Each microinequity contributes to the overall message that a student is unwelcome.

We all say microinequities when we speak to others. Most of us will never achieve a level of social and cultural awareness that allows us to avoid *all* such language. As teachers, our words carry a little more weight. We are almost never in casual conversation with students because of our position of authority.

With practice, you will start to recognize and avoid microinequities. Better yet, you can promote inclusion by incorporating "microaffirmations" when you talk with students. Microaffirmations are small statements that let students know they are welcome and capable. These definitions are adapted from the National Alliance for Partnership in Equity:

**Microinequity**—Small statements (usually unintentional) that send a message that students are unwelcome or incapable.

**Microaffirmation**—Small statements (usually intentional) that send a message that students are welcome and capable.

Consider the following classroom scenarios. Each is based on a statement made by a well-meaning TA during a classroom observation. Test yourself! In each case, try to identify the microinequity in the statement and consider how the TA could have worded things differently to make the statement more affirming.

### SCENARIO 1

**TA:** "What is the purpose of tube #2 in this experiment?"

**Student:** "That tube has the substrate but no enzyme so we can make sure nothing happens."

**TA:** "OK. What I think he was saying is that it is important to include a *negative control* when performing this experiment."

*The microinequity:* The student is fundamentally correct but did not use a term the TA is looking for (negative control). "What I think he was saying" is condescending and undervalues the student's useful contribution to the discussion.

*Make it a microaffirmation:* "Great! That's exactly why we include sample #2 in the experiment. We have a term for this." At this point, the TA could follow up with the same student, ask another student, or ask the whole class for the term, "negative control."

> **SCENARIO 2**
>
> **TA to the class:** "Guys! Guys! I need your attention up here! You guys aren't being careful enough when measuring out the enzyme buffer. If you push the pipette plunger down all the way, you will draw up about twice as much as you need!"

*The microinequity:* Not everyone in the class is a "guy." In fact, fewer than 50% are men if this class follows national trends. Even though referring to a mixed-gender group as "guys" is common, it sends an exclusionary message to some students.

*Make it a microaffirmation:* Replace "guys" with "students" or "everybody." You can use something more discipline-specific like "biologists" or "statisticians" if you like the way that sounds.

> **SCENARIO 3**
>
> **TA:** "So a catalyst—it's like that one person on your floor of the dorms, when nothing is happening and it's really quiet, who goes around and gets everybody to watch a movie or go out to eat or something. They are the catalyst!"

*The microinequity:* The TA used an analogy to help students understand a concept—great! The choice of analogy is unfortunate. For many students, dorm life is not part of their college experience, and this analogy could make them feel excluded.

*Make it a microaffirmation:* The same analogy could easily be applied to a more universal social environment. Alternatively, the TA could say "Sometimes you hear someone in the news described as a 'catalyst for change.' What is the reporter telling us about this person?"

You will not always get it right—none of us will. Keep reminding the students and yourself, "I want everyone to feel like they belong here. I want everyone to feel encouraged and successful in this class." Do your best to promote these affirmative feelings and reflect on your words and interactions.

## EXCLUSIONARY LANGUAGE FROM STUDENTS

You can regulate your own words in the classroom, but what about the language *students* use? Students may say things that are insensitive or exclusionary during class discussions or casual classroom conversations. To maintain a classroom environment where everyone feels included, you may find yourself in the position of having to push back against microinequities said by students. It is uncomfortable, but do not ignore these situations and learn to respond calmly and with compassion for the speaker and for their peers. Research suggests that simply trying to silence the speaker ("You can't say that!") is ineffective and that the alternative approaches below are more effective in addressing exclusionary language by students (adapted from Samuels 2014):

*Use questions:* "Why do you say that?" or "Do you feel that way about every person in that group?"
*Tell the speaker how you feel:* "It makes me uncomfortable to hear you say that."
*Suggest that the words are not aligned with the speaker's personality:* "I'm surprised to hear you say that because you seem like a very thoughtful person."

---

**BOX 1. PRONOUN PRESSURE**

Asking students to share their names and personal pronouns has become a common start-of-class practice. Each student indicates if they identify as she/her, he/him, they/them, ze/hir, or something else. Sharing pronouns began as a sincere desire to signal that the classroom was a safe space where student identity was respected. It seemed like a positive step toward inclusion. Sharing pronouns was also a response to the fact that class rosters typically show legal names that transgender or nonbinary students might no longer identify with.

Starting a class with sharing personal pronouns might not be as welcoming and inclusive as hoped (Levin 2018). Students can be reluctant—or terrified—to share their gender publicly, on the first day of class, to a room full of strangers. One outcome is that social pressure leads a student to resort to the most common pronouns (she/her or he/him) when they identify as something else. You can find guidance on pronoun sharing at mypronouns.org. Here are some quick recommendations:

**Encourage students to share pronouns during introductions, but do not make it required.** Say, "Please introduce yourself and tell us your pronouns if you wish."
**Refer to all students by their names.** Rather than "Did everyone hear what he said?" say, "Did everyone hear what Jake said?" (Don't know all the names yet? Just ask. "Great idea! Tell me your name, please.")
**Include your own personal pronouns** in your introduction and on the syllabus and follow that up with "I want to refer to everyone correctly, so please consider sharing your own personal pronouns with me privately, or by e-mail."
**Have each student make a name card** for their desk or table and encourage them to list their personal pronouns. You will need to provide materials (notecards) for this. Ask students to save the name cards for future classes or collect them yourself and hang on to them for the next class meeting.
**Pronouns change!** Invite students to change their pronouns at any time.
**If you are meeting with a student in office hours, ask.** "My pronouns are [your pronouns here]. How should I refer to you?"

More colleges and universities are allowing students to list their preferred names and pronouns in the class roster system. Check to see if your institution does this and encourage students to update their profiles.

## DO A PRETERM SURVEY

Earlier in this chapter we described an inclusive teacher as someone who recognizes diversity in their classes and takes intentional actions to make the educational experiences equitable for all students. This was followed by a list of possible actions that you could apply in your teaching. In this section, we will examine a way to recognize the diversity among the students in your class.

Some diversity is clearly visible. But some very important forms of diversity are invisible. Characteristics like learning style, degree of introversion or extroversion, and mental health status are not readily visible but will have a big influence on how a student experiences your class. You will not know about these forms of diversity unless a student volunteers the information or you ask and really listen to the responses.

A tool you can use to gather some of this information is a preterm survey, a short anonymous survey that students complete before the term starts or during the first weeks. There are many benefits to doing a preterm survey. It allows you to introduce yourself and the course and sends a message that you care about getting to know who the students are. The biggest benefit is that it allows students to describe themselves to you anonymously and voluntarily. This gives you a much more complete view of the diversity within your class. Addy et al. (2021) developed the "Who's in Class" form, an excellent preterm survey that is freely available online (see resources at the end of this chapter). Their work inspired the survey example below:

- **Instructions to students**: "This survey will help me know better how to support all students in [class name here]. The survey is anonymous and individual answers won't be shared. You aren't required to answer every question. Thank you!"
- Why is it important for you to do well in this class? How does it fit into your future plans?
- What would you like me to know about you as a learner?
- What factors outside of class could affect your ability to be successful?
- **At the end:** "If you would like to discuss any of your answers here with me, please get in touch. My e-mail is... ."

A pre-term survey should be anonymous. Anonymity will encourage openness in the responses. The survey should help you identify the range of learning styles in your class, letting you take intentional actions to support all learners. For example, if multiple students report that cold calling (randomly calling on students to speak in class) is very stressful, you can use alternative forms of participation (see Chapter 6). You can say to the class, "In the preterm survey, several of you said the idea of getting called on in class made you anxious, so rather than call on individuals, I'm going to ask you all to spend one minute writing your ideas about the topic."

There are easy ways to create and deliver a preterm survey. Platforms like Google have free survey tools. Your school or department may have a license for commercial survey software. The campus learning management system that runs your course web page (Canvas, Moodle, or others) will have a survey tool built in. If your campus does not have specific guidelines about what survey platform to use, pick whatever you are most familiar with. Of course, you do not need technology at all. Write a paper survey, hand it out in class, and give the students 5–10 minutes to complete it during your initial class meeting.

If you are reading this a few weeks into the term, it is not too late! You will need to change the wording of the introduction and some questions to account for the timing. But it is never too late to get to know your students better and to make changes that improve inclusivity. You could also use this survey to get feedback on your teaching so far. Consider including questions like, "I'm eager to improve as an instructor. Do you have any suggestions for me?" or "How is my lecture style working for you? Do you have any suggestions that would make it better?" You are making yourself a little vulnerable by asking questions like this. In my experience, new TAs who have asked for this sort of feedback midway through the term have found the student responses to be mostly thoughtful, respectful, and helpful. But be prepared to hear about some things that you have no control over, like "Class shouldn't start at 8 am … it's too early!"

### WHAT DOES AN INCLUSIVE CLASSROOM LOOK LIKE?

"A CLASSROOM WHERE EVERYONE FEELS LIKE THEY CAN EXPLORE THIS NEW MATERIAL AND BE THEMSELVES AUTHENTICALLY WITHOUT FEELING UNSAFE. WHETHER THAT HAS TO DO WITH THEIR IDENTITIES OR THEIR LEARNING ABILITIES—EVERY STUDENT FEELS SAFE TO BE BOLD."
— Katie, 2nd year graduate student

"IT'S EASY TO FOCUS ON THE PEOPLE WHO ARE MORE TALKATIVE OR OUTGOING. OR YOU'RE MORE LIKELY TO ENGAGE WITH STUDENTS THAT ARE LIKE YOU RACIALLY OR CULTURALLY. BUT IN THAT KIND OF SITUATION—THAT KIND OF BIAS—IT'S UP TO YOU TO SEE THE PEOPLE THAT ARE SHY OR NOT TALKING AND HELP THEM ENGAGE."
— Jeremy, 2nd year graduate student

"I'VE TRIED TO NOT JUST BRING LESSONS ABOUT [TOPIC], BUT LESSONS ABOUT LIFE BECAUSE THEY'RE ALL STUDENTS, AND WE ALL GO THROUGH STRUGGLES. TAKING A STEP OUTSIDE OF THE CLASSROOM AND SAYING, 'WE'RE ALL PEOPLE. I'M HERE FOR YOU. IF YOU NEED RESOURCES. I CAN HELP YOU GET RESOURCES.' I THINK THIS OPENS THE CLASSROOM A LOT AND STUDENTS FEEL A LOT MORE COMFORTABLE."
— Annie, 2nd year graduate student

## Want to Know More?

Addy TM, Dube D, Mitchell KA, SoRelle ME. 2021. *What inclusive instructors do: principles and practices for excellence in college teaching.* Stylus Publishing, London.

> *The authors interviewed more than 300 teachers from diverse institutions and distilled the information down to clear messages about how these teachers define inclusivity and how they support diverse students in their classes.*

Addy TM, Mitchell KA, Dube D. 2021. A tool to advance inclusive teaching efforts: the "Who's in Class?" form. *J Microbiol Biol Educ* **22**: e00183–21.

> *This publication includes the complete text of the "Who's in Class" form—a preterm survey instructors can use to learn about the characteristics of their students as people and as learners.*

Dewsbury B, Brame CJ. 2019. Inclusive teaching. *CBE Life Sci Educ* **18**: 2.

> *This paper links to an online evidence-based teaching guide featuring a large bibliography of research about best practices in inclusive teaching. It is a great way to browse a lot of research to find resources to improve your teaching.*

Tanner KD. 2013. Structure matters: twenty-one teaching strategies to promote student engagement and cultivate classroom equity. *CBE Life Sci Educ* **12**: 322–331.

> *A review of twenty-one proven strategies that improve classroom equity.*

thesafezoneproject.com/resources/courses/

> *SafeZone Training—Online courses for instructors who want to learn the best ways to support LBGTQ+ students.*

CHAPTER 6

# Participation: Promote Active Learning

Once new TAs have a few weeks of teaching experience and start to put first-day nervousness behind them, many wonder, "How do I get the students to talk more?" or "How can I increase student participation?" They imagine a classroom where students are asking and answering questions, thinking critically about the material, and discussing ideas. These are all components of what educators call "active learning." Active learning happens when students *do* things and *think* about what they are doing (Bonwell and Eisen 1991). The knowledge students gain through active learning is more enduring because they have a role in constructing their understanding of new topics (Mintzes and Walter 2020). When students are active participants in their learning, student understanding goes up and failure rates go down (Luckie et al. 2012; Freeman et al. 2014).

But you already know this. To learn something new and challenging, you need to *do* stuff: Draw it out, explain it to a partner, predict what comes next, and test yourself. As a teacher, you can make these types of activities a regular part of your classes. Not only will students learn better while in class, but they will also incorporate these active strategies into their studying improving their overall academic skills. Creating opportunities for active learning is a way of teaching that you develop with practice and reflection.

Teaching and learning can be broadly divided into two categories—passive and active—and each has advantages and disadvantages (see table). Distinguishing between passive and active learning will help you make good choices about the best way to prepare for and teach a class. The answer to the question, "Who's talking?" is a way to identify a passive or active classroom. With passive learning, the teacher does most of the talking and the students mostly take notes. During active learning, the students are given time to think about the new ideas they are learning and frequently talk about these ideas. The role of the TA in the more active classroom is to guide thinking by asking good questions and to facilitate discussions to ensure that many students have a chance to talk.

The default for many new TAs is a teacher-centered style in which students are passive learners. One reason is that new TAs are replicating the type of instruction they are used to seeing. You have almost certainly been in a traditional lecture-style class in which the instructor (the "sage on the stage") was the source of information

|  | PASSIVE | ACTIVE |
| --- | --- | --- |
| Is described as… | teacher-centered | student-centered |
| Is best for… | conveying a lot of information that students need to know or remember. A lot of content is covered quickly. | helping students make connections and think critically about the topic. |
| Who is talking? | The teacher does most of the talking; students talk infrequently. | Students talk frequently about course-related ideas. |
| What are students doing? | Students listen and take notes. | Students solve problems and discuss ideas. |
| What is the instructor doing? | Mostly lecturing and demonstrating. | Some lecturing and demonstrating, but mostly guiding students and facilitating discussions. |
| Preparation required by an instructor | Focused on content (What will you present and how will you present it?). | Focused on content and on student engagement (What needs to be presented so students can start to work independently or in groups?). |

and your role was to take notes. For many new TAs, their expectations for themselves as teachers have been defined by the teaching they have been exposed to, and they believe "That's how it's done." These TAs may work hard to make their classes clear, organized, and engaging, but understandably, they reproduce the teacher-centered instruction they experienced.

Another reason new TAs default to a style in which students are more passive is that the preparation required for passive teaching is somewhat less. The easiest class to prepare is one where you do all the talking. You only need to think about what you are going to say and you have complete control over that.

The underlying idea of this chapter is that including some active learning in each class session is a good goal. With active learning, students become more engaged with the subject of the class, and they learn better and faster. That said, this chapter is not a condemnation of a passive teaching style. Sometimes there is a lot of information that needs to be handed off to students, and a passive, teacher-centered model may be the best way to do that. There will be times

"I DON'T KNOW... IT JUST FEELS VERY SAGE-ON-THE-STAGE-Y."

when you need to fall back onto a more teacher-centered style simply because the material is unfamiliar to you or your time for preparation is limited. It is OK to say to yourself, "This time, I'll do most of the talking, but in the future, I will look for ways to make it more interactive."

If you teach a laboratory class in which students do hands-on work in groups, you might be surprised to learn that even in this format, active learning isn't a certainty. A student can follow the steps of a lab protocol without thinking critically about what they are doing and why. Even though they are doing things with their hands, they may be a passive participant. As a lab instructor, you have a role to play in promoting active learning.

Before we look at specific ways to make active learning happen, let's look at two classroom scenarios. One takes place in a lecture setting and one in a lab setting. The common theme is that the TAs initially instruct in a teacher-centered way with the students as passive learners. Then the TAs make changes that incorporate active learning.

### ACTIVE LEARNING IN A LECTURE: IDEAL GAS LAW

Elena is a TA leading a discussion section for an introductory chemistry class for non-science majors. It is her first time teaching the class. Elena and the other TAs were told in the coordination meeting that the goal of this week's sessions is for students to review the ideal gas law and do practice problems.

"In lecture you were introduced to the ideal gas law." Elena writes "$PV = nRT$" on the board. "$P$ stands for pressure, $V$ is the volume, $n$ is the number of molecules in moles, $T$ is the temperature, and $R$ is a constant." The students write down the equation and take notes as Elena talks.

"The equation shows the relationship between all these terms. Pressure and volume have an inverse relationship. As one goes up, the other goes down. So, if you increase the pressure, the volume goes down. Think of a balloon. A balloon is filled with gas. If you decrease the pressure around the balloon, and keep everything else the same, the balloon gets bigger. Volume goes up." Elena continues this way until she has mentioned all the variables in the equation. The class transitions to doing calculations using the equation.

As you read the description of Elena's presentation to the class, you probably recognized moments when Elena could have made the students more active participants.

Elena reflects on this session and wants to make it more interactive. In her Teaching Reflection she writes, "Ask students to use their everyday knowledge to come up with some of the relationships." She adds some comments in the margins of her teaching notes with ideas for questions she can ask.

It is the next semester and Elena is preparing to teach the same class again. She reads her Teaching Reflection and spends some time thinking about how to encourage more active participation in the review of the ideal gas law. It's worth noting that when teaching the ideal gas law for the second time, Elena has had a full semester of experience teaching chemistry for nonscience majors. She's more comfortable with public speaking. She is more familiar with the content of the class and how it ties together. She knows what to expect in terms of background knowledge from her students. Her experience makes the transition in her teaching from passive to active much easier.

Elena begins, "Think of a balloon. It's filled with gas. There's pressure inside and outside of the balloon. If the pressure outside the balloon goes down, how will the balloon change? You can use your intuition, or you can use the ideal gas law equation you learned in class."

After a few seconds of silence, Elena says, "Everybody, how will the volume of the balloon change? Thumbs up if you think it will get bigger, thumbs down if you think it will get smaller. Good, almost everyone is putting thumbs up. Now imagine a graph with pressure on one axis and volume on the other." Elena draws the $x$- and $y$-axes on the board. "What would this graph look like? Take 30 seconds to sketch it out."

After 30 seconds, Elena says, "Compare your graph to someone sitting near you. Did you sketch the same thing?" Elena briefly leaves the front of the room and glances at some of the graphs the students have drawn.

Returning to the front of the room, Elena says, "Carla, did your table agree on what the graph should look like? Could you describe it?" Carla uses a hand motion to indicate the direction of the line on the graph and Elena draws it on the board.

"Putting this all together," Elena says, "we are seeing the inverse relationship between pressure and volume. That's reflected in the ideal gas law. Your graphs show that as pressure goes down, volume goes up." Elena writes "$PV = nRT$" on the board. "Now let's think about the other variables in the equation and make predictions about what will happen when they change.

Elena found simple ways for the students to be more active. Using a thumbs up/thumbs down question, she invited the entire class to participate rather than just one student. She asked the students to build on their initial answer by drawing a line on a graph showing the pressure–volume relationship. Elena created an opportunity for the students to talk about the graph they drew. By the time Elena drew the equation on the board, the students had actively thought about the relationship between two of the variables.

## ACTIVE LEARNING IN A LAB: NEGATIVE CONTROLS

Leo is a new TA for an introductory cell biology class for Biology majors. The subject for this week's lab is enzyme activity and how it is affected by factors like temperature and substrate concentration. In the coordination meeting, the TAs were reminded to focus on the use of controls in the activity because experimental controls are a core concept the students would use throughout the term.

Leo begins, "You should have already read the steps for today's procedure. Open to page 25. In step 4 you are each going to prepare two tubes. Notice that they are the same except for one thing. One of them has the enzyme substrate and the other replaces the substrate with water. That second tube is our negative control."

Leo's presentation has been entirely teacher-centered, and the students have been passive. He communicated clearly and the students were likely prepared to start working quickly. However, Leo missed an opportunity for the students to think critically about the lab and to practice communicating with each other.

Jump to the following semester. It is time for the enzyme lab again. Leo uses his experience to make the session more active. He starts, "Open to page 25. Something important is happening in step 4. Take 30 seconds at your table to say out loud how the two tubes you prepare will differ." As the students talk, Leo leaves the front and walks around the edge of the room listening to the conversations.

Returning to the front, he says, "Great. I was listening and I can tell everyone knows what they are going to do. Now let's think about when this lab is over, and you're getting ready to write your report. Can you explain why we need that second tube with the water in it? How is it working for us? Try to explain that to another person in your group."

Leo has asked a harder question, and it takes a little longer for the groups to start talking. Leo moves around the room encouraging the groups, saying "This is a hard thing to put into words. See if your table can come up with an explanation together."

Back at the front, Leo wraps it up. "Sam, I heard you use the correct term when talking about tube 2. Could you repeat that? Right, it's the 'negative control.' Someone from this table in the back, why do we need a negative control in this test?" Leo continues to ask for student input until he feels they arrive at a complete explanation of the role of negative controls.

Leo worked in two opportunities for active participation. He still drew the students' attention to the importance of step 4 of the protocol. Rather than tell the students the what and why of the step, Leo started by asking the students to explain *what* they were going to do. This was quick, and by listening to the student conversations, Leo decided no further discussion was necessary. The students were then asked to articulate *why* they were including the control. To give the discussion more urgency, Leo reminded them that they would need to know this for their graded lab reports. Leo facilitated the discussion so multiple students could share their thoughts.

These two scenarios illustrate that it does not have to be difficult to switch from a teacher-centered style to a more student-centered one. The same information was covered using both modes. The TAs used a variety of techniques to get the students to talk about ideas and to get feedback from the students. The strategies below will help you incorporate more active learning in your teaching.

## Ask Good Questions

Asking good questions is key to initiating active learning in your classroom. The way a question is worded can either encourage or discourage student participation. No TA sets out to intentionally ask questions in a way that discourages participation—it is more likely that they have not thought about what they are trying to achieve with the question or the way the question will be perceived by students. With a bit of thought outside of class, we can ask questions in a way that encourages all students to participate. With practice, it becomes second nature.

Not all questions are equal in terms of how much thinking they require to answer. Just because you are asking many questions during class does not assure active learning. It is helpful to have a system for identifying the type of question you are asking and why you are asking it. One such system divides the questions teachers ask into four classifications: rhetorical, managerial, open, and closed (Blosser 1991). These classifications are a useful starting point.

**Rhetorical questions** emphasize a point or introduce a topic. For example, "Have you ever wondered why the periodic table is organized the way it is?" or "Can you picture a world where intricate ecosystem webs do not exist, and each organism stands alone in isolation?" These are the least interactive types of questions. The teacher may not even want or expect a response. Use rhetorical questions if that is your style, but make sure they are not the only type of questions you ask.

**Managerial questions** keep the class moving. We ask managerial questions to check if students are prepared to advance to the next thing. In a laboratory setting, these may be very practical ("Does everyone have a square of cheesecloth and four test tubes?"). The most important managerial question in a lecture or lab can be simplified to, "Are you confused?" Knowing whether students are confused about a topic or procedure is important for knowing if the class is ready to move on. The simple question, "Are you confused?" is unlikely to provide meaningful feedback because the wording implies that the confusion is the student's fault. No one enjoys feeling confused. Publicly admitting you are confused in front of your peers makes the feeling worse, especially for students who may already be experiencing imposter feelings. Careful wording of managerial questions can alleviate the self-consciousness or peer pressure that discourages students from answering. Here are alternatives to "Are you confused?" that get a better response:

- "Is there anything that requires more explanation?"
- "Have I explained this well enough?"
- "Are there any questions before we begin?"
- "What do you need to know before we move on?"

**Open questions** ask, "What do you think?" With open questions, a wide range of responses are acceptable, rather than just one correct answer. Open questions ask students to draw on their knowledge, experience, opinions, and values and apply those to concepts important for the class. These are the sorts of questions that scientists and academics ask themselves which lead to great insights or discoveries. Examples of open questions include, "Was the invention of plastics a good thing or a bad thing?" and "You have a family history for a genetic disorder and a DNA test becomes available—would you take the test?" These are great questions to begin a discussion and to get students thinking about important ideas.

**Closed questions** are the most common types used in classrooms, especially math and science classrooms. Closed questions have a limited number of acceptable or correct responses and often just a single correct response (the "right answer"). In other words, most closed questions can be summarized as "What's the answer?" Because they are so common, we will take a closer look into the problems and opportunities that closed questions create.

Closed questions signal to students that there is an all-or-nothing correct answer. They perceive that the TA is looking for a very specific response. The author Bryan Brown (2019) has a good name for this: the "Guess-What's-in-My-Head game." The stakes feel high for answering this sort of question—getting the wrong answer could make the student look bad. Students will be reluctant to answer such questions unless they are very confident in their abilities and are pretty sure they've "got it." The rest of the students, the ones who are building their understanding of the concept, may be reluctant to participate—and may even mentally opt out. But it is these students, not the students who are already confident in their knowledge, who would benefit the most from thinking critically and trying to answer.

It does not have to be that way. You can ask closed questions in ways that make the stakes feel lower and that encourage more students to think and try to answer. Let's compare high-stakes and lower-stakes versions of closed questions from a biology class:

| HIGH STAKES | LOWER STAKES |
|---|---|
| "What does 'catabolism' mean?" | "When a cell is in a catabolic state, what's changing in the cell? What types of processes are occurring?" |

The high-stakes question is a big ask—to articulate a definition of a core concept. The lower-stakes version guides students to think about what is important. The lower-stakes version also sends the message that multiple different answers could apply and that more than one student can participate in answering.

| HIGH STAKES | LOWER STAKES |
|---|---|
| "Is Parent 1 in this cross homozygous or heterozygous?" | "Parent 1 in this cross is heterozygous. What evidence do we have that helps us determine that?" |

To lower the stakes, the TA has given away the answer (the genotype of the parent). Consider how giving the answer away changes the dynamic of the question and the classroom. Students who had the right answer in their heads are empowered to share the thinking that got them there. Students who had the wrong answer in their head got the message privately—without having to be wrong in front of the group. (On a future exam, the students may be expected to answer the high-stakes question. But during class time when they are still learning how to think about genetics, the lower-stakes version will better serve most of the students.)

In addition to the four classifications of questions described above (rhetorical, managerial, open, closed), there is another category that TAs can use to help their students learn actively: the **How-Do-You-Know** question. These types of questions help students think critically about the knowledge they are building and the problem-solving strategies they are employing. How-Do-You-Know questions use wording like, "Why do you think that?" or "What were the steps?" This type of question has elements of both open and closed questions. Like open questions, they help students feel like their knowledge and experience are relevant and valuable. Like closed questions, they may have a single correct answer. After all, there are usually only a small number of ways to solve a problem, and sometimes only one way.

Incorporating How-Do-You-Know questions into teaching sends a message that the process, not the answer itself, is the most important thing. How-do-you-know questions ask students to explain their thinking, which is valuable to students who know—or guessed at—the right answer, and to their peers who are still looking for ways to think about a topic or solve a type of problem.

### Wait Time

You gaze out at the students. The question you just asked hangs in the air. The clock ticks. Some of the students stare back at you while others avoid the discomfort of eye contact. Time drags on... .

This scenario sounds unpleasant. No one likes an awkward silence. It feels *especially* awkward if you are the one in front of the class. Ten seconds of silence can feel like forever. However, silence can be a tool for getting students to participate. Learn to embrace the silence and make it work for you and the students.

Returning to the scenario described at the start of this section: It is a few seconds later and there is still nothing from the class. You take a deep breath, smile, look expectantly out at the class, and let a few more seconds pass. A hand goes up and you respond to the student. "Sarah, what do you think?" Then, "Thank you. That's a really good idea. Does anyone have another idea?" Another hand or two goes up and you have a discussion going. The wait, awkward as it was, was productive.

"TAKE THREE MINUTES TO WORK ON THIS PROBLEM—EVEN THOUGH I'LL STOP YOU AFTER 30 SECONDS BECAUSE THE SILENCE IS TOO AWKWARD."

Good questions asked in class, questions that help students learn, should require students to think before they answer. Even though the wait time may seem very long to you, avoid the temptation to cut it short. (Some teachers sing a verse of a song in their head or recite the alphabet to themselves as an internal timing mechanism.)

An often-cited series of studies in the 1970s (reviewed in Tanner 2009) revealed that, on average, teachers waited just *1 second* between posing a question and calling on a student to answer. A modest increase in wait time—from 1 second to 5 seconds—resulted in more students participating, longer, more complete answers, and more correct answers.

Sometimes you might ask a very basic question that does not require any thought and gets an immediate answer. Managerial questions like "When is the lab report due?" or "How many milliliters of salt solution do we add at step 5?" fit this description. There is nothing wrong with minimal wait time when asking questions like this. But active discussions require the kinds of questions that need some time for thought.

A very long silence could indicate that the students really *do not* know how to answer. They might not have enough background knowledge to answer your question. Another explanation could be that your question was confusing or unclear. If the silence drags on, it is fine to say, "Let me ask the question in a different way," or "Let's take a step back," and take another run at it. Simplifying the language, providing some background information, or giving a hint could make answering the question possible.

Another strategy to address a too-long silence would be to initiate a brief Think-Pair-Share around the topic (this strategy is discussed in Chapter 7). Students might be reluctant to answer in front of the whole group but willing to say their ideas out loud in a smaller peer group.

## Hand Signals

Answers spoken out loud are not the only way students can participate during class. For many questions, you can ask the students to use their hands to display answers. The advantages of this approach are that it is quick and creates an opportunity for every student to answer. Hand signals work best for managerial or closed questions with yes/no, true/false, or "choose one" answers. Recall Elena's lecture described earlier in this chapter. She asks the entire class to participate: "Everybody! Thumbs up if you think it will get bigger, thumbs down if you think it will get smaller." If you use this approach, you will find it easy to scan the room and get a consensus that you can report back to the students: "OK, I see most of you putting thumbs up. That's right!" If you are asking for hand signals, tell the students to wait and display their signals at the same time: "Keep your hands down to give everyone a chance to think about it. OK, hands up in 3, 2, 1…go!" Here are more examples:

- "Is the force greater in this direction or in that direction?"
- "Look at the image on the screen. Which arrow is pointing to a hydrogen bond?" (arrows are labeled 1, 2, 3, and 4)
- "How confident are you that you could explain this concept to someone else? Hold up fingers with 5 being 'totally confident' and 1 being 'not confident'."

There are electronic systems (clickers or phone apps) that can accept student input. These can store, report, and display response data. The output from these platforms can also be linked to a gradebook to make in-class questions graded for accuracy or participation. Hand signals represent a less robust form of whole-class participation but without some of the hassles and technical problems of an electronic system.

## NORMALIZING ERROR

How you respond when students participate will influence future participation. If students feel *successful and appreciated*, they will continue to participate. If students feel *unsuccessful and unappreciated*, they will stop participating.

What about when the student who participates is wrong or off track? You must show that you value their participation while also correcting the response. Making mistakes is part of learning. If the teacher is dismissive of students who make a mistake, they disrupt learning. Your words and attitude should send the message that mistakes are OK and a natural part of learning. Teachers refer to this as "normalizing error." When a student gives a wrong answer, you should accept that answer and use it as an opportunity to promote learning.

It is important not to excuse mistakes. It can be tempting to say, "That's right!" when you should really say, "That's *mostly* right," or "You're off to a good start." Do not

Participation: Promote Active Learning    69

let your desire to make students feel supported lead to you signing off on an answer that is only partially correct.

As you gain teaching experience, two things will happen that will make it easier to address a wrong answer: (1) you will develop language that will help you normalize error and (2) you will have a response ready for the common mistakes and misconceptions that come up every term. Until then, here are some phrases you can use to turn student mistakes into teachable moments. As the chart below shows, you can choose to ask the student who participated to follow up on their answer, open the discussion up to the rest of the class, or explain the answer yourself.

Here are more suggestions for normalizing error and correcting incorrect—or partially incorrect—responses:

"**I can see why you would think that** because these two concepts do *seem* very similar."

"**I really like what you said**, and you are on the right track. But let's think about how we can push this idea further."

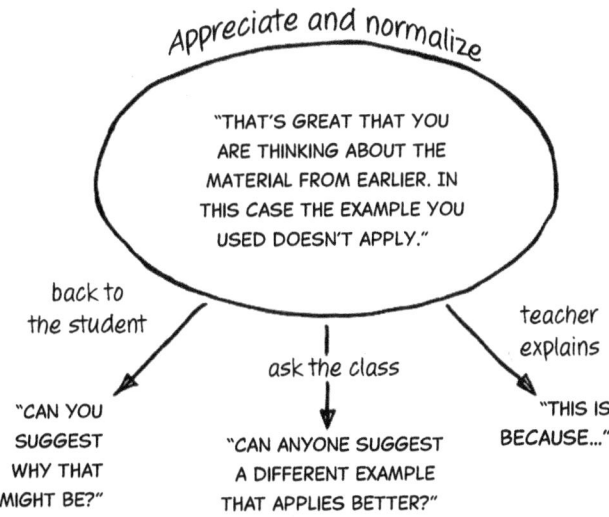

"**I'm *really* glad you said that,** and I bet a lot of other students are thinking the same thing. This is a very common source of confusion."

"**Great! You have the *concept* completely right.** However, I want to suggest a wording change that could make the statement more accurate."

"**Can you restate that** using class vocabulary?" or "Can you restate that but incorporate the word, [vocabulary word here]?"

"**Can anyone add anything** that could make this answer more complete?"

## BE MOBILE

"I'M MR. SOCKY! YOUR TA WILL BE STAYING BEHIND THE INSTRUCTOR DESK TODAY."

You are familiar with the typical set-up of college classrooms: desks in rows and a podium in the front near the chalkboard and projector screen. A laboratory classroom has fixed tables with seating for four to six students and an imposing instructor bench up front. These layouts send a message even before class has started—the instructor has a stage on which to perform, and the students are positioned to watch the show. In other words, traditional college classrooms are not designed for interaction. There are times when you need to deliver important information in a traditional lecture style and this layout is helpful. However, if your goal is to encourage students to be active, you need to find a way to "get off the stage."

The easiest thing to move in the classroom is *yourself*. As with many things in teaching, this will take some getting used to. It can be comforting to stand behind a podium or instructor table and be physically separated from the class. This is especially true early in the term when the students are strangers to you. Moving out from behind the podium and into the student space might feel a little uncomfortable, but breaking that imaginary barrier is a simple way to show that you see the classroom as a shared learning space where everyone participates.

In some classrooms, the student desks might be movable. Maybe not *designed* to move, but not bolted to the floor. Small group discussions will be a little easier if you instruct the students to "Rotate your desks to form groups of three or four" or whatever group size you have chosen (groups of three or four work best for most activities). Ask students to return their desks to the original position at the end of the class. The instructor following you will appreciate that you have reset the room.

Some newer classrooms and laboratories have been designed for active learning with seats that can swivel to facilitate group discussion and labs set up to encourage collaboration. Trying to teach in a traditional, teacher-centric way in a setting like this will be tough. If you find yourself teaching in a classroom designed for active learning, do your best to incorporate some of the techniques described in this chapter and Chapter 7.

## GETTING EXCELLENT PARTICIPATION—FROM ONE STUDENT

The focus of this chapter has been on encouraging active learning by all the students in your class. How do you handle a situation in which there are one or two students who enthusiastically participate, and how does their participation influence the rest of the class? Having an enthusiastic participant is *not* a bad thing. You might be happy to know you have a student who will break the uncomfortable silence after you ask a question. However, you and the other students in the class can become dependent on that student, and that limits more broad participation. Students can quickly adopt the expectation that when you pose a question to the class, they do not have to engage because that student in the front row is always going to volunteer to answer and you are always going to call on them. The instructor has a role in either promoting or dismantling this dynamic.

There are polite ways to handle this scenario. You can show that you value that student's contributions while also encouraging others to participate. Here are some suggestions:

- *"I have a question for the class. Don't raise your hand if you know the answer. I want everyone to think about this for a moment."* Prefacing your question this way prevents a hand from going up right away. When that hand shoots up, brains shut down—why bother to think about the question when another student already knows the answer? After asking your questions and waiting, say, "Who would like to suggest an answer?"

- *"We've already heard from Jay; I'd like to hear from someone else."* Jay is a student who is always eager to answer your questions. You quickly acknowledge Jay's willingness to answer, then shift your attention to the rest of the class. Avoid the temptation to return to Jay if you don't get a quick response from another student. Returning to Jay would only reinforce the idea that students can get by without active engagement.

- *"I'd like to hear from someone who hasn't said anything yet today."* This is another way to let students know that you are looking for participation from new voices. If the same hand or hands are raised in response to a question, and approaches like those above have not worked, consider initiating a Think–Pair–Share (see Chapter 7).

Consider reaching out to the enthusiastic student after class or via e-mail to say, "I really appreciate that you are such a good participant in class, thank you." Then explain your *educational* reasons for why you chose to call on others. The student will understand.

## COLD CALLING

STUDENTS QUICKLY REALIZED THAT COLD CALLING WAS A BIG PART OF PROFESSOR P'S TEACHING STYLE.

Some instructors do not ask for raised hands, and use the class roster to call on students. This approach—cold calling—results in more students participating. However, cold calling puts a little pressure on students. For some students, this pressure is productive leading them to think more critically about the question than they might have otherwise. For other students, the pressure of knowing they might be called on is counterproductive, triggering anxiety that overwhelms the capacity to think about the question.

There is evidence that cold calling, especially during the first weeks of a term, can have benefits. One study compared two sections of the same science class. In one, cold calling was used extensively early in the term. The students in the high-cold-call section showed greater *voluntary* participation later in the term than the students in the no-cold-call section. This increase in voluntary participation was especially significant for students who identified as female. Moreover, neither women nor men reported greater discomfort in the high-cold-call environment (Dallimore et al. 2019). So cold calling could be a route to greater and more equitable participation. If you choose to use cold calling in your class here are some guidelines (adapted from Becker et al. 2018):

*Lead with cold calling:* Make cold calling your primary way of getting student feedback rather than resorting to cold calling if there are no volunteers.

*Inform the class and follow through:* At the start of class say, "I will be calling on every single one of you today," and then be sure to do that.

*Have a system to keep track of whom you called on:* Some teachers write each student's name on a card and pull from the deck. Others write each student's name on a popsicle stick and pull the sticks out one at a time.

*Give them time:* Make sure you provide a wait time long enough to allow students to process the questions. Let the students know, "I'm going to call on a few students to share their thinking in 30 seconds."

### Other Forms of Participation

This chapter has focused on a specific type of participation: asking or answering questions during a discussion involving the whole class. There will always be some students who are reluctant to participate in this situation, no matter what approaches the instructor uses. It is simply not their style. If your goal is for students to be active learners there are forms of in-class participation that *don't* involve speaking in front of the class. For students who are reluctant to participate during class discussions, try these approaches:

*Check-in with individuals:* When the class is working in groups or doing lab work, ask the student individually, "Did you have any thoughts about that question I asked earlier? Did you agree with what was said?"

*Prepare students to participate:* During a discussion, listen in (or join in) with small groups of students who typically do not participate. If they are having a productive discussion, ask them if they would be willing to share their idea with the class. This could lower some of the barriers to participating in the class discussion because they already know their contribution will be useful. (This approach is sometimes referred to as "warm calling," an alternative to the cold call.)

*Do one-minute notes:* Rather than ask for hands to be raised, pass out paper—maybe a quarter sheet per student—and tell the class, "Write your answer on this paper or, if you aren't sure of the answer, write a question you have about this topic. You have one minute." This requires planning and materials. If your class is small, you can sort through the responses you receive and get a flavor of what students are thinking. You can pick out a few at random and share them anonymously or you could intentionally pick the responses of students who typically do not volunteer and share those—again, anonymously—so those students hear their words as part of the discussion. (Note: In the recent past you could instruct students, "Tear a blank page out of your notebook," and have them use that for in-class writing assignments. With more students using laptops and tablets as their primary note-taking devices, they may not have paper. Anticipate this either by bringing small pieces of paper with you or by allowing paper-free students to e-mail their responses to you.)

*Be attentive:* A student who is reluctant to speak in front of the class may be an active participant in small-group discussions, during lab work, or during other types of activities. This student is engaging in active learning. The fact that the student does not readily volunteer to speak during whole-class discussions is not a concern.

### Want to Know More?

Blosser PE. 1991. *How to ask the right questions.* National Science Teachers Association, Arlington, VA.

> *A booklet describing the different question types with classroom examples of each. The author explains how to recognize and use each type of question. Using silence and wait time are also discussed.*

Lemov D. 2021. Teach like a champion 3.0: 63 techniques that put students on the path to college. John Wiley, Hoboken, NJ.

*Short chapters each focusing on a different teaching technique presented as a classroom scenario (many with links to videos of teachers in action in their classrooms). The audience for this book is K–12 teachers, but many of the techniques translate well to college classrooms.*

Mintzes JJ, Walter EM, eds. 2020. *Active learning in college science: the case for evidence-based practice.* Springer, New York.

*A collection of articles describing different techniques to promote active learning. Evidence for the effectiveness of each technique is presented.*

Tanner KD. 2009. Talking to learn: why biology students should be talking in classrooms and how to make it happen. *CBE Life Sci Educ* 8: 89–94.

*This article reviews the evidence that student talk promotes learning. The author then examines the reasons why some instructors do not include student talk (it takes up useful class time, it is impractical with a large class, etc.) and makes suggestions to overcome these perceived barriers.*

CHAPTER 7

# Think–Pair–Share: Get Them Thinking and Talking

THINK–PAIR–SHARE (TPS) IS a tool instructors can use to promote discussion and help students learn. For many instructors, TPS is an intentional part of every session. The reason is straightforward: TPS provides a format for students to think about and talk about important concepts. When done well, TPS can make the class more inclusive and give a greater number of students a voice by promoting discussion and lowering the participation barrier. As you will see, in a TPS activity, you can call on a student to share their *group's* opinion, rather than their *individual* opinion. This makes participating a little less scary.

From the teacher's standpoint, there are several advantages to TPS. TPS offers a fairly easy way to shift lecture focus away from yourself during those times when you seem to be doing all the talking. In other words, TPS can help you transition the class from passive learning to active learning. Another advantage of using TPS is that it gives you the opportunity to get out from behind the podium and get a sense of how the students are expressing their ideas. Finally, there is evidence that activities like TPS, which promote participation early in the term, lead to more participation by more students later in the term (Dallimore et al. 2019).

If there is one teaching technique to become comfortable using, Think–Pair–Share is it. Like so many other elements of teaching, TPS must be planned and practiced, and this chapter will guide you as you plan TPS activities for your class. Once you have more experience, you will find it easy to implement TPS spontaneously when you sense that the students need some time to think about a hard concept in a lecture or lab. Once you gain experience at running TPS in your class, you will find that it has a central place in your teaching.

## OVERVIEW OF A THINK–PAIR–SHARE ACTIVITY

**THINK**
- Give the students a task with a well-defined product (e.g., make a list, write a sentence that explains a concept, or solve a problem).
- Tell the students how long they have to generate the product.
- Say "No talking for now."
- Students work quietly and independently.

**PAIR**
- Remind students of the product and tell them how long they have to work.
- Say "Discuss your [list, solution, etc.]. Spend some time trying to resolve any differences. Try to come up with a complete [list, solution, etc.] for your group."
- In small groups or by table, students compare their products from the Think phase.
- Suggest that one group member consolidates ideas from all group member's products.

**SHARE**
- Instructor calls on a group or Individuals to share part or all of their product.
- This can happen fast. The most important part of the activity has already happened.
- Before closing ask, "Does anyone have anything else to add?"

### Think

You have been in class when the instructor has asked a question, then followed it with, "Turn and discuss this with your neighbor." Your mind goes blank as you realize you need to start talking, perhaps with a stranger, about a potentially new and difficult concept that you have had almost no time to think about. You make eye contact with your neighbor who is probably thinking the same thing. The only students who feel comfortable in this scenario either (a) already know the answer or (b) are sitting next to a friend. Students who are new to the material or who do not have social connections in the class are at a disadvantage. The instructor may congratulate themselves that they have initiated a discussion; however, there is no guarantee that the discussion will be productive.

For a discussion to be productive, students must be given a clear mental task and a reasonable amount of time to think about it before they talk. They will do so knowing that after a short time, they will share what they are thinking to at least one other student—this gives urgency to the thinking. Your instructions to the class must be very deliberate at this point. Here is an example of an in-class review of photosynthesis:

> "Photosynthesis starts with a photon of light hitting a leaf and ends with sugar (glucose) being produced. Between those two forms of energy, there are several intermediate forms. Try to list as many of the intermediate forms of energy as you can. Try to list them in the order in which they occur. Please work on this on your own for one minute with no talking for now. If you are stuck, try to think of a question that will help you make progress. After one minute you'll have a chance to compare your list with other students around you."

This carefully planned introduction includes a well-defined product (a list or a question), a duration (one minute), a reminder that this is individual work ("no talking for now"), and anticipation of the sharing step coming up next.

One minute can feel like a long time to stand in front of a quiet classroom. Avoid the temptation to end the Think phase early. If you said, "one minute," give the students a full 60 seconds. Even if some students appear to finish after 30 seconds, there will be others who use the whole time.

With experience, you will develop an intuition about how much time to give the students. You can always add time. After the given time, if you see many students still working say, "That's one minute. Who needs more time? I see a few hands so keep working individually for another 30 seconds." It is likely that some students will use more time than others. To acknowledge this, you can add, "If you are at a stopping point, think through it again. See if you can add anything else or refine your initial ideas." Or "In your head, practice how you will explain this to another person."

### Pair

The second phase of the TPS activity is very important. The Pair phase provides a time and a format for students to have an *academic* discussion. "Saying it out loud" helps students practice using class vocabulary. But beyond vocabulary, talking with a peer helps students make personal connections with their classmates, helps them see classmates as allies in the learning process, and gives them a low-stakes way to evaluate their understanding (Kelly 2019). If they really struggle to explain their thinking to another student, they may realize what concepts or details need more attention.

Do not put too much weight on the term "pair" when thinking about group size. Depending on your class and the topic, groups of two, three, or four are fine. Larger groups might work as well. For example, in a laboratory classroom with six students around each lab bench, it might make sense for the whole bench to work as a group. But, in general, two to four is the most productive group size.

Back to the photosynthesis example:

"We are thinking about all the transfer of energy during photosynthesis. Everyone should have one or more forms of energy listed. In groups of two or three, take two minutes to compare your lists. Your goal is to work together to come up with the most complete list possible. Give everyone a chance to talk. I'll walk around and answer questions."

The TA has done several important things to help the students get started on this phase of the activity. They restated the goal of the activity, indicated the expected group size (two to three) and work time (two minutes), and redefined the product (a consolidated list from the group). By saying, "Give everyone a chance to talk," the TA is providing some guidance about how the students should interact in their groups.

Stay active and involved during this phase of the activity. Now is not the time to take a two-minute break! Leave the front of the classroom and walk around. This is your opportunity to take the temperature of the class. Listen in on groups. Are many students contributing, or just a few? Are they using class vocabulary correctly? Being out among the students will encourage them to ask questions. You will be surprised at how much more willing students are to ask you questions when there is lots of talking in the room and you are not up in front of the class. A student who would feel reluctant or self-conscious to interrupt a lecture to ask a question might do so readily during the TPS. Answer procedural questions quickly ("How much time do we have for this?"), but if a group has conceptual questions, take more time with them. Rather than provide an answer immediately, respond in a way that helps the group get "unstuck" and promotes further discussion. Try pointing the group toward the resource they need to answer: "Michael, I see you have the lab protocol up on your screen. Scroll up to Figure 1. Tina, what do you notice that might help your group answer your questions?"

Sometimes, a group will not feel like they have anything to discuss. Let us say you have asked the students to work on a task like calculating a chemical concentration or using a formula to solve a physics problem—something where there is a correct answer. A group might compare answers, see that they all have the same thing written down, and then sit quietly thinking they have nothing to talk about. A TA can anticipate this by announcing at the start of the Pair phase of the activity, "If everyone in your group of three has the same answer, use the time to share your work and see if you can explain your solution to each other." If you notice a group sitting quietly during, you can ask "Did you all get the same answer? Good. Try to explain your thinking to each other." You can also prompt that group with a follow-up question like, "How would your answer change if the [starting concentration, pH, gravity] was lower?"

You want noise at this point. A noisy classroom indicates that the students are making progress and learning. When the noise level starts to drop, it may be time to move on. If you have overestimated the time needed, you can bring it to a close early. You can say, "That's been a little over a minute and I'm not hearing much talking. Raise your hand if your group is at a stopping point."

### Share

During the Share phase of TPS, the instructor asks one or more of the groups to share the results of their discussion. Sharing is important for bringing closure to the activity. Also, the Share lets you put your stamp of approval on the product you have asked for. Finally, students might have thought of new questions about the material now that they have had more time to actively think about it.

The Share phase can be quick. As you listened to discussions during the Pair phase, you may have noticed that the general level of understanding was high. That's great! Do a quick Share and move on. Alternatively, if there wasn't much conversation happening during the Pair, expect to spend a little more time on the Share. In this case, you can use the observations you made as you checked with the Pair groups. If you were asked the same question multiple times, answer it for the group. You can always direct the students back into their pairs for more discussion after you have given the group some guidance.

The Share phase is also when you can encourage participation by students who don't often use their voices in class. When you call on someone to speak, make it clear that you are asking them to be a spokesperson for their group ("Ava, could you tell me something your group discussed?" versus "Ava, how would you answer this question?"). This can make participation seem less scary. Let's wrap up the photosynthesis example:

> "It looks like most groups have reached a stopping point so I want to move on. Thank you for working on this. Group 1, what's the first thing on your list? 'Excited electron.' Good! Group 2, what do you have next on your list? 'A proton gradient.' That's right." (TA writes responses on the board to create a class list.)

The TA has signaled the end of the Pair phase and redirected attention back to the front of the class. The TA thanked the class for being active and made a class product—a list of photosynthesis energy forms in this case.

This is the time to identify and address misconceptions. In the case of the photosynthesis example, a group may have suggested something that *should not* be on the list. Say, "I'm glad you said this because it is a common misconception." Address it and move on.

Do your best to get as many of the groups to contribute to the Share. In the case of our photosynthesis example, you could ask each group to contribute one item to the list until the list is complete. If you have a large class or many small groups, you might not be able to ask every group to share something. That's OK. Before you wrap up the activity ask, "Does anyone have anything else they want to add? Did any new questions come up during your discussion?" This opens the door to individuals with thoughts or questions to contribute an idea or question they did not discuss during the Pair phase.

With planning and practice, TPS will become a useful part of your teaching toolbox. Initially, you will spend time outside of class planning TPS activities. When you have more experience, and your students are used to having TPS as part of each class session, you can initiate TPS "on the fly" when you sense that students need some time to process new information or when you want to incorporate a break from a more lecture-oriented material. Once you are familiar and comfortable with the basics of TPS, you will find that the approach can be easily modified to suit whatever you are doing in class.

### Think-Pair-Share Planner

With experience, you will be able to run TPS activities with minimal planning. Initially, it's good to plan out each phase. Fill in the attached planning worksheet if it's helpful or just use the prompts as a guide. (You can download a copy of the worksheet at TheTAsGuide.com.) A sample worksheet is included on the next page.

**What's the product?**
At the end of this activity, what should the students have generated? In the space below, write or draw what you consider to be a complete product.

---

**THINK** (Time:        )
Write the question/prompt you will say to the class so students can start working toward the product.

Remember to say, "No talking yet" or "Work independently for now."

---

**PAIR** (Time:        ) (Group size:        )
Write the question/prompt you will say to the class so groups can continue to work toward the product.

---

**SHARE** (Consider alternatives to a whole-class share)
Write how you will conduct the *Share* phase of the activity.

Remember to ask, "Do any groups have questions?" and "Does anyone have anything to add?"

**What's the product?**
At the end of this activity, what should the students have generated? In the space below, write or draw what you consider to be a complete product.

Pedigree on pg. 10

Students write the genotype of each person

Students calculate the probability that '?' will be aa.

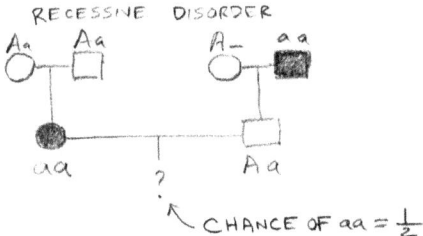

**THINK** (Time: 2 min )
Write the question/prompt you will say to the class so students can start working toward the product.

Look at the pedigree on page 10. Try to write in the genotype of each person and calculate the probability that the unborn child (the question mark) will have the disorder.

Remember to say, "No talking yet" or "Work independently for now."

**PAIR** (Time: 3 min. ) (Group size: 2–3 )
Write the question/prompt you will say to the class so groups can continue to work toward the product.

- In groups of 2–3, compare pedigrees and probabilities.
- Goal is to collaborate to come up with a 'high confidence' answer in your group.
- If you have differences, try to talk them out. If everyone has the same thing, practice explaining to each other how you know the genotypes of the individuals.

**SHARE** (Consider alternatives to a whole-class share)
Write how you will conduct the *Share* phase of the activity.

- Draw pedigree on the board.
- Ask one group to explain how we know the disorder is recessive. (Use A and a for the alleles.)
- Ask 6 different groups to each give the genotype of one person on the pedigree (I write on board).
- Ask 1 group for the probability (I write on board).

Remember to ask, "Do any groups have questions?" and "Does anyone have anything to add?"

## VARIATIONS OF TPS

There are ways to alter and adapt the basic TPS format described in this chapter. Instructors are free to explore variations and alternatives to the standard TPS to discover what works best for you and the students in your class. Here are some possible variations adapted from Cooper et al. (2021):

*Consent to Share:* During the Pair phase discussions, the instructor listens in. Upon hearing a group discuss an important idea, the instructor asks a group member, "Would you be willing to share that with the class in a few minutes?" It is likely that the student will consent to speak, but even if they don't,

"NOW WE WILL DO A THINK-PAIR-SHARE-PAIR-SHARE-THINK-SHARE-PAIR-SHARE!"

the instructor has sent a message to that student saying, "I hear and value your ideas." This strategy might be especially helpful if your whole-class discussions tend to be dominated by a few voices—you can get new voices into the mix.

*Instructor Summary:* The instructor listens in during the Pair phase. Rather than invite students to speak, the instructor summarizes important points acknowledging, if possible, where they heard those ideas. For example, "During the discussion, Manuel said something very important, which was that chromosome number stays constant throughout mitosis. Did that come up in any other discussions? [Nods from around the classroom]" This approach might be best when you sense that most of the class has a good understanding of the material or if you need to wrap up quickly.

*Delayed Sharing:* This requires a little extra preparation. During the Think and Pair phases, the instructor passes out notecards or quarter-sheets of paper. After the Pair phase, each student writes their name and ideas on the paper and hands them in. The next class session begins with the instructor sharing some of their favorite ideas, giving credit to each student. "Many of you wrote something along these lines, but Tracy summarized it very nicely. She said… ." This method could also be done electronically using a survey app. Just be ready to provide a paper option for students who cannot, for whatever reason, access the survey in class.

*Repeat the Pair:* Talking about hard concepts is shown to promote learning and improve the use of vocabulary used in your discipline. If you value having students talk about ideas in class, why not create more opportunities for talk by skipping the Share phase in which only a few students speak. Instead, initiate a new Pair phase. "Now I'd like you to find a new partner (or group) and state your idea again. See if your new partner has a similar idea or something different. Remember, 'saying it out loud' helps you learn, so try to use this opportunity to practice using class vocabulary." This approach gives students a chance to move around and interact with new people in the class.

Regardless of whether you use the standard TPS described in the first part of this chapter or one of the alternatives described above, try to incorporate at least one TPS in each class session. Soon, TPS will start to feel like a natural extension of your teaching. As you prepare for an upcoming class, look for opportunities to integrate TPS. As you reflect on a class you've already taught, think about whether there was anything you lectured about that could have worked as a TPS—and make sure you write about this in your weekly reflection. If you provide students with the time and the format to talk about hard concepts, you will start to see the general level of discussion and participation increase over time.

### Want to Know More?

Cooper KM, Schinske JN, Tanner KD. 2021. Reconsidering the share of a Think-Pair-Share: emerging limitations, alternatives, and opportunities for research. *CBE Life Sci Educ* **20**: fe1.

*A critical evaluation of Think-Pair-Share. The authors suggest that teachers and students value TPS differently. Several interesting alternatives to TPS are proposed that might overcome some of the shortcomings of TPS.*

CHAPTER 8

# Grading: Make it Fair and Efficient

EXAMS, QUIZZES, LAB REPORTS, AND ESSAYS—these are all examples of assessments used to determine what students know and what they can do. The focus of this chapter is on creating and grading these types of assignments, collectively called formal assessments. Even when you are not giving a quiz or exam, you are still assessing students. Informal assessment, the moment-to-moment feedback you get from observing students and listening to their discussions, is also very useful when teaching. But for now, we are going to be formal and focus on graded activities.

Few instructors—and almost no students—relish the task of assessment. But assessment is necessary. Students must be evaluated and graded.[1] A challenge with assessment is creating tools to accurately measure student knowledge and learning. Another challenge is evaluating students without bias. Every student must be evaluated fairly whether their quiz ends up on the top or the bottom of the grading stack.

Assessment is a time-management challenge. Many new TAs underestimate the time required for grading. Part of the time management challenge is that the grading workload is not equally distributed across the term. The amount to grade ramps up as the term progresses. Not coincidentally, you may find yourself grading midterms while also studying for your own exams. For all these reasons, trying to be as efficient as possible is important. Fortunately, it is possible to make assessments that effectively

---

[1] Some individual instructors and even a few institutions are experimenting with systems in which instructor grading is reduced or eliminated (sometimes referred to as *ungrading*). Students work with their instructors to define their learning goals, reflect on their progress toward those goals, and ultimately assign themselves a grade (Flaherty 2019).

evaluate what students know, and that do not require a large time commitment to grade. As with anything in teaching, assessment gets easier with experience.

This chapter focuses on two types of assessment for which TAs are likely to have the most responsibility and autonomy: in-class quizzes and lab reports. The discussion of quiz writing will help you create assessments that test student understanding at multiple levels so you can distinguish the best-prepared students from the rest. For grading essays and lab reports, this chapter will give you ways to make the process more efficient and more fair.

## *GIVING* THEM A GRADE

Before jumping into the details of writing and grading assessments, let us consider the appropriate mindset when it comes to evaluating students. New TAs often say how much they hate "giving a student a bad grade." Most TAs were the type of student who wanted to perform well in class, and grades were an obvious way to judge their performance. You probably felt good when you got a high grade and bad when you got a lower grade. It is natural for you to impose some of your own feelings on students in your class. You want those students to be successful, and it is easy for that impulse to creep into your mind while grading. Avoid this! It opens the door for bias and lowered expectations.

At the end of the day, you are entering a letter grade into a spreadsheet. Try not to think of it as you *giving* the student a grade. You provided guidance for students as they learned the material. You reached out to students with academic challenges. You did your best to write fair assessments to gauge student learning (and this chapter will help with that). Some students performed well on these assessments, and some performed poorly. You used information from assessments to determine what grade the student *earned*. Thinking back to yourself as a student may be helpful. You knew that a low grade was a result of being underprepared. You took it as a signal that you needed to work harder or change how you were studying—and you did not externalize blame on your instructor.

## BLOOM'S TAXONOMY

Not all quiz questions assess the same sort of thinking and learning. Say you want to determine whether students understand an important concept in metabolism: how the movement of electrons is coupled to ATP synthesis in mitochondria. Asking students to list the proteins of the electron transport chain in order is not an effective way to do that. You would be testing whether they *memorized* the names and not whether they *understand* how the system works. Memorizing is one type of cognitive task. Explaining a complex system is a different cognitive task. The questions we ask of our students, particularly in quizzes and tests, are not the same in terms of the cognitive tasks they represent.

Bloom's taxonomy is a system used to classify educational tasks based on their complexity and the type of thinking involved. The taxonomy was first described by Benjamin Bloom and his colleagues in the 1950s, and it has been revised many times since then. Bloom's taxonomy gives us a framework for thinking about the questions we are asking and why we are asking them. As you write a quiz, ask yourself, "What am I trying to assess?" and "Does this question really assess that thing?" Bloom's taxonomy provides a structure for this critical thought. The modern version of Bloom's taxonomy has six levels: *remember, understand, apply, analyze, evaluate,* and *create* (Anderson et al. 2014). Any quiz question you write will fall into one of these levels. A well-written quiz or exam should include questions from multiple levels.

| BLOOM'S LEVEL | QUESTIONS AT THIS LEVEL ASK STUDENTS TO… |
|---|---|
| Remember | Define the term… <br> Name the parts… <br> List the steps… |
| Understand | Compare two types of… <br> Describe the process… <br> Explain the relationship between… |
| Apply | Calculate the value… <br> Predict the outcome… <br> Explain how a system will change in response to… |
| Analyze | Analyze a data set to determine… <br> Examine a microscope image to distinguish… <br> Interpret the graph… |
| Evaluate | Agree or disagree with the conclusions… <br> Explain which method would be best to determine… <br> Choose the best treatment and justify your selection… |
| Create | Design an experiment to… <br> Write an alternative hypothesis… <br> Develop a plan… |

Each Bloom's level requires a different type of thinking. In a genetics class, being able to *define* the term "genotype" is important. Being able to *determine* the genotype of an organism by analyzing data from breeding experiments is also important. The first task assesses scientific vocabulary and sits squarely at the *Remember* level of the taxonomy. The second task assesses the ability of the student to think like a geneticist and corresponds to the *Analyze* level of the taxonomy. To properly grade students in a genetics class, you need to know how well they can do both tasks.

Below are examples of test items at every level of Bloom's taxonomy. The format for all of them is short answer or fill-in-the-blank. Later in this chapter, you will see how some of these could be rewritten as easy-to-grade multiple-choice questions.

**Remember:** To what class and order do octopi belong?

**Understand:** What adaptations enable octopi to efficiently extract oxygen from water?

**Apply:** Predict how the range of the Pacific octopus will change as ocean temperatures increase.

**Analyze:** Examine data from physiology experiments to determine the relationship between water temperature and octopus hunting activity.

**Evaluate:** Critique current conservation efforts intended to protect octopus populations worldwide.

**Create:** Outline a marine protection plan for octopi that accounts for their life cycle.

What Bloom levels would you guess are mostly used by novice assessment writers? When new TAs start writing assessments, most of the questions tend to be at the *Remember* or *Understand* level of Bloom's taxonomy. There is nothing wrong with these questions—a good assessment will have questions at these Bloom's levels. However, to be able to distinguish the best-prepared students in your class from the less-prepared students your assessment should also include questions at other levels of the taxonomy. By asking questions at multiple Bloom's levels, especially those that include high-order thinking (*Apply, Analyze, Evaluate*), you will be able to assign grades that reflect the varying levels of understanding likely to be found in a typical class. If you, like other new instructors, gravitate toward the *Remember* and *Understand* levels in your quiz writing, recognize that tendency, and remind yourself to write questions at other levels.

## "BLOOM" YOUR QUIZZES

Some TAs write their own assessments. Others are provided with a bank of quiz questions to choose from. For large classes with multiple sections, the quizzes may be standardized, and the TAs have little or no opportunity to modify the questions.

Whatever your situation or assessment type, "Blooming" your quiz will be useful. Recognize the Bloom's level of each question. If possible, add or modify questions so they fall into different Bloom's level. If you are working from a pre-existing test bank, try to recognize the question level and select a good variety. Blooming your quiz in this way ensures a better assessment.

Even if you are not involved in writing the assessment, understanding the Bloom's level of the questions will help you to help students. A student who consistently misses *Remember* questions needs to modify their studying differently from a student who struggles with *Evaluate* questions. The first student needs to devote time to memorization (perhaps with flashcards or a digital equivalent) and the second needs to practice explaining complex ideas and predicting outcomes. You can help students develop important academic skills by recognizing the Bloom's level of your assessment questions and sharing that knowledge.

## WRITING AN ASSESSMENT

If you are responsible for writing and grading assessments for your class, this section will show you how. If you are not the primary assessment writer for your class, you will gain insights into the process that will inform your teaching and make you a better guide for the students you teach—and probably help you become a better test taker.

Before you start writing questions, get organized. Jumping straight into writing assessment "items" (this term refers to questions, essay prompts, and other tasks) can result in wasted effort because the items you write might not target the most important ideas, or they may be at the wrong cognitive level. The first thing to do is to identify the big ideas, important details, and essential skills the students should understand and be able to do. After teaching a discussion or lab section, you can probably identify these core topics on your own. If you want help, the course textbook or lab manual will often list the learning objectives associated with each topic.[2] Whether you write it yourself or find it in course materials, this list of objectives is a time-saving guide for assessment writing.

Next, consider the appropriate Bloom's level for each topic on your list. If interpreting graphs or analyzing data is expected of the students, questions at the *Analyze* level must be included to assess how well they can do that. If memorizing names, dates, or structures is important for the course, you will need to assess that at the *Remember* level. At the end of a term, you might want to see if students can connect ideas from across the

---

[2] The use of learning objectives (LOs) in teaching is an important topic for anyone developing a course. Creating LOs that articulate what students are expected to know, understand, and do should precede the creation of lectures, labs, schedules, and exams. Everything students are asked to do as part of the class should help them achieve the LOs. (For an overview, see Orr et al. 2022.)

entire course to *Evaluate* a plan or *Create* an experiment. Identifying the Bloom's level required for each learning outcome prepares you to write assessment items efficiently.

Finally, consider the format and length.[3] If you are creating a short quiz to make students accountable for assigned reading, a few multiple-choice questions might be appropriate. If you are creating an end-of-term exam intended to sum up many weeks of learning, you will probably need to include a variety of question types including some that require a written response. If you are teaching one section of a large-enrollment course, there may be a required format and length that all instructors follow. Your own time constraints should also factor in here! Do you have time to grade multiple written responses this week? If not, your time might be best spent writing some good multiple-choice items.

## Write Multiple-Choice Items

Compared to other easy-to-grade question types, multiple-choice items offer several advantages. Multiple-choice items can be used to assess students at different Bloom's taxonomy levels. Although not the best for assessing student's ability to *Create*, multiple-choice questions can be written to assess how well students *Remember*, *Understand*, *Apply*, *Analyze*, and *Evaluate*. An advantage of multiple choice over True/False is that multiple-choice items are less susceptible to guessing, making multiple choice a more reliable gauge of student understanding. Because of these advantages, and because they are quick to grade, it is worthwhile to learn to write good multiple-choice items.

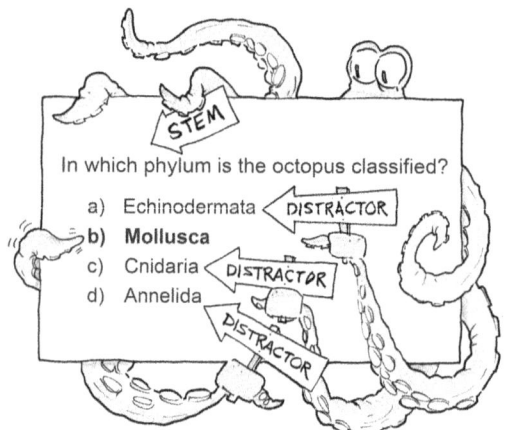

A multiple-choice item has two parts: the stem and the alternatives. The stem is where the instructor poses the problem to be solved. The alternatives are the solutions the students choose from. These include the correct or best choice (the answer) and the incorrect or inferior choices (called the "distractors"). Here is advice for writing stems and distractors (adapted from Brame 2013 and from Burton et al. 1991):

***The best stems pose a question:*** When possible, write multiple-choice stems that present students with a clear, complete question. A good way to check for this is to ask yourself whether a student could read the stem and formulate an answer without needing to look at the answer choices. For example, a stem that asks, "Which

---

[3] Want an idea of how much time it will take students to complete your assessment? Time how long it takes you to read and answer every question. Multiply your time by four or five to get an estimate of how long it will take the students. If it takes you three minutes to complete your quiz, expect that students will need 12–15 minutes.

statement about gravitational force is correct?" falls short. The stem, "On which factor does the strength of the gravitational force between two objects depend?" is better. (Answer: Gravitational force depends on the distance between the objects and their masses.)

*Stems can be partial sentences:* A fill-in-the-blank type question can be useful for assessing vocabulary. If you use a partial sentence, try to put the blank toward the end. The stem. "The \_\_\_\_ is the plant organelle primarily associated with photosynthesis," would be better if the blank were at the other end: "The plant organelle primarily associated with photosynthesis is the \_\_\_." Putting the blank at the end of the stem allows the student to focus on identifying the best answer rather than on the structure of the sentence. (Answer: chloroplast.)

*Avoid negative phrasing in the stem:* When possible, avoid negative phrasing in the stem, as this can confuse students and make the question less reliable in grading. Students who understand the larger concept might miss the questions because they get turned around by the negative wording. Avoid wording like "Which of the following is not true…" or "What isn't a characteristic of…" in the stem. (If there is no way to avoid using negative phrasing, format the negative words so they are hard to miss: "Which of the following *IS NOT* an approved safety measure in CHEM 101?")

*Distractors should be plausible answers:* When writing distractors, your goal is to write choices that will be difficult for less-prepared students to distinguish from the correct choice. To make the distractors seem like plausible answers, make them about the same length or slightly longer than the correct answer. Incorporate class vocabulary in the distractors—though the vocabulary may be used incorrectly or inaccurately. As you gain teaching experience, you will become more aware of common misconceptions and student errors. Incorporate these misconceptions and errors in the distractors. If a student calls you over during a quiz or test and says, "I feel like there is more than one correct answer," this indicates that you did a good job with the distractors. Assure the student, "There is just one correct answer," and leave it at that.

*The number of distractors can vary:* Do not put pressure on yourself to come up with the same number of distractors for every question. If the distractors are plausible options, having as few as two distractors is fine.

*Avoid grammatical clues that give away the correct answer:* When writing fill-in-the-blank questions, avoid grammatical choices that signal the correct answer. In the example below, there is only one choice (a, the correct choice) that completes the sentence in a way that is grammatically correct:

In the experiment from last week's lab, temperature was an \_\_\_\_.

(a) **independent variable**     (c) negative control

(b) dependent variable     (d) positive control

There is a simple fix to this problem: move the article out of the stem and into the answer choices:

In the experiment from last week's lab, temperature was ____.
(a) **an independent variable**  (c) a negative control
(b) a dependent variable  (d) a positive control

### Write Free-Response or Short-Answer Items

Asking students to write, draw, or show their work as they solve a problem is an excellent form of assessment and the best way to assess thinking at the higher levels of Bloom's taxonomy: *Apply, Analyze, Evaluate, and Create.* Free-response and short-answer questions will make grading more challenging in terms of time and consistency. Here are some suggestions for writing these types of questions that can reduce these challenges.

***Take Time to Save Time:*** There is an old saying: "Ten hours of grading can save you ten minutes of quiz writing." Maybe that is not really an old saying, but the point is that items that are quick to write often take a long time to grade. Spending a little extra time writing a prompt for a free-response item will pay off when you sit down to grade a large stack of quizzes. The biggest problem to avoid is vague wording in the prompt. If you are vague in your question, students will try to cover all the bases leading to long, wordy responses. For example, if your goal is to assess the students' understanding of experimental design, you could quickly write, "What are experimental controls?" Quiz written! But when you sit down to grade, be prepared to search through lots of long, wandering responses as you look for evidence that students understand the concept.

With a little extra effort, you can write a prompt that gives the students more guidance and lends itself to responses that are brief, to the point, and faster to grade. Present students with an experimental setup—maybe like something they have seen in a lecture or lab—and follow up with, "Identify the controls in this experiment. Explain how each control allows the researcher to interpret the results." Compared with responses to the quick-to-write version of this item, student responses will be **easier to compare** because students will all use the same source material (the experiment you describe). The second prompt gives **specific guidance** about the information you want students to provide. This makes it easier for you to create a key for quick and accurate grading (more about grading keys later). The wording of the prompt provides students with a way to **organize the response** (identify control, explain, identify control, explain, etc.). More organized responses allow for faster grading.

Do not expect quiz writing to be quick. But keep in mind that any extra time you invest in writing the assessment will allow you to save time during grading.

***Limit how much students write:*** You can write assessments in such a way that even free-response questions have built-in constraints that limit how much students write. For example, you can give guidance about the expected length ("Answer in one or two sentences"). You can provide students with lines on which to write (see below) with specific instructions:

"Answer using the space below. Write one idea per line. You are not required to use every line in your answer."

_____

_____

_____

_____

(It is easy to make lines like this in your word processing program. Create a table that is one column wide and with as many rows as needed (four rows in the example). The height of each row in should be 0.3–0.4 inches (0.75–1 cm) to give students sufficient space to write. Finally, format the table to remove the vertical borders on the edges.)

***Could that short-answer question be multiple choice?*** Asking students to express their ideas in writing is an excellent form of assessment. However, it is not necessary for students to provide written answers for *every* question. Consider the following free-response item:

"Explain the relationship between alleles and proteins."

This can be answered in one well-organized sentence. Most students will probably write more. Consider whether the same relationship could be assessed with a multiple-choice item:

Which statement below best describes the relationship between **alleles** and **proteins**?

(a) **Two alleles of a gene encode distinct proteins or patterns of gene expression.**
(b) Proteins make up the alleles of genes.
(c) Mutations of proteins create new alleles.
(d) Identical genes can have multiple alleles of proteins.
(e) Two genes differ by what alleles are encoded in their proteins.

In general, asking students to choose "the best" answer is a way to increase the difficulty of the questions because the students are being asked to discriminate between choices that all have some degree of correctness (Carriveau 2023). Asking students to choose the best statement takes a question from the *Remember* level of Bloom's taxonomy to the more challenging *Understand* level. In the example, choice A is the best answer that articulates the relationship using the correct vocabulary. The distractors, alternatives B through E, all have something that makes them inferior to A—incorrect use of vocabulary or conceptual problems. As discussed earlier, it is important to make sure that the distractors are plausible options. In this case, the distractors are similar in length and tone to the best answer. Students must understand the relationship between the terms and think critically about how each choice is articulated. Obviously, the multiple-choice version is much quicker to grade than the free-response question we started with.

## Crafting a Chemistry Quiz

Elsie is a TA for a laboratory section of *Introduction to Physical Chemistry*. Before each lab, the students are expected to think critically about the concepts and results of the previous week's lab and prepare for this week's lab by reading the background and procedures in the lab manual. It is Elsie's job to write and grade a quiz that will assess these expectations.

This will be a busy week for Elsie. She has a presentation for the Graduate Chemistry Seminar on Thursday and the initial draft of her thesis research plan is due to her committee by Friday. It would be beneficial to reduce the time needed for grading this week. With time constraints in mind, Elsie decides to include only one item that requires a written answer. The rest will be multiple choice.

Last week's lab was about oxygen and combustion. Students generated $O_2$ from hydrogen peroxide and then burned wood and other materials in low $O_2$ and high $O_2$ environments. Elsie will write two questions about this lab. The questions should be at the *Apply* or *Analyze* level of Bloom's taxonomy. These levels are appropriate because the students have done the lab and, with Elsie's guidance, thought critically about the procedures and results.

This week's lab is about acids, bases, and pH. Questions about this lab should be primarily at the *Remember* or *Understand* levels. It is premature to ask students to analyze data or evaluate conclusions because they have not done the lab, collected data, or discussed results. The main goal of these questions is to determine whether the students prepared for this week's activity, and questions at the lower cognitive level of Bloom's taxonomy are appropriate.

Elsie reviews her teaching notes and the lab manual. She identifies four topics she wants to assess with the quiz and writes them down:

> **From last week:** (1) Explain the role of the catalyst in the production of $O_2$, (2) compare combustion in a low $O_2$ and a high $O_2$ environment.
>
> **For this week:** (3) Review the relationship between proton concentration and pH, (4) know the procedure for using a pH probe.
>
> Elsie is ready to start writing test items. For the first topic on her list, she writes: What was the purpose of the catalyst in the reaction used to generate $O_2$ from $H_2O_2$?

This question is fine, but in the interest of saving time during grading, Elsie converts it to multiple choice. She spends time writing distractors that are plausible and that echo some of the misconceptions that students expressed about catalysis during the in-class discussion:

> What was the primary role of the catalyst in the chemical reaction that generated $O_2$?
>
> (a) To increase the temperature of the reaction.
> (b) To increase the number of $O_2$ molecules produced from each molecule of reactant.
> (c) To decompose resulting in the formation of $O_2$ and $H_2$ gas.
> **(d) To increase the rate at which the reactant decomposed to form $O_2$.**

For the second item on her list, Elsie is interested in assessing how well a student can apply what they learned to a new situation. Because she will be assessing higher-order thinking, Elsie decides to use her one free-response item. She thinks about what the students did last week and uses her creativity to come up with a new scenario that involves some of the same procedures and concepts.

> A lab group is testing combustion of different papers. They observe that printer paper burns slowly and produces little light while tissue paper burns quickly and produces a bright light. Each student suggests an explanation for what they observe:
>
> **Alan:** "The tissue paper has more carbon than printer paper so it makes more $CO_2$."
> **Beto:** "The tissue paper has bigger holes so there's more contact with the air."
> **Cara:** "Combustion is faster in the tissue paper so it's a better catalyst than printer paper."
>
> **Who do you agree with the most? Support your answer with evidence from last week's lab.**
> _____
> _____
> _____
> _____

This question is at the *Apply* level of Bloom's taxonomy. By asking, "Who do you agree with the most?" Elsie is assessing whether students can use their knowledge to discern between explanations. Additionally, students must make connections between the paper-burning scenario and their actual observations from the previous week. The lines Elsie added below the question give students clues to the expected length of their answer. (Note: Beto's explanation is the best choice.)

Writing items for this week's lab is a little easier. With the lab manual open Elsie scans the material the students were assigned to review prior to lab. Her goal is to ask straightforward questions that any well-prepared student could easily answer. Here are Elsie's questions for objectives three and four on her topic list:

*How does an acidic solution differ from a neutral solution in terms of proton concentration and pH?*

*(a) Greater proton concentration, and lower pH*

*(b) Greater proton concentration, and higher pH*

*(c) Lower proton concentration, and lower pH*

*(d) Lower proton concentration, and higher pH*

*Correct use of the pH meter requires which of the following procedures?*

*(a) Submerging the probe up to the handle in the solution you are testing.*

*(b) Rinsing the probe only after all measurements are complete.*

*(c) Ensuring the tip of the probe does not contact the glass of the test tube.*

Elsie's quiz writing is done. She saved time by identifying the concepts she would assess and considering the Bloom's level appropriate for each before she started writing questions. She made conscious choices about the question format that took her busy schedule into account. Finally, she took her time writing the quiz to ensure that the questions were precise and that student writing would be limited.

## FAIR, EFFICIENT GRADING

### Use a Key to Grade Written Responses

Asking students to articulate their understanding in writing is a good form of assessment. Free-response or short-answer questions give you insights into students' thinking and let you assess whether the students can use the class vocabulary.

Written responses present several challenges for grading. Maintaining consistency as you work through a big stack of quizzes can be difficult. Quizzes on the bottom of the pile must be analyzed with the same rigor and criteria that you apply to quizzes on the top. Another challenge of grading written work is that it is subjective. You must put guardrails around the grading process to maintain fairness and eliminate bias as much as possible. (See Box 1: The Halo Effect for more about how instructor bias can affect grading.)

---

**BOX 1. THE HALO EFFECT**

In an unbiased grading system, students who submit similar work should get similar scores. But unbiased grading is not a given (Malouff et al. 2013). Psychology instructors at multiple institutions watched a video of a student doing an oral presentation. One group of instructors saw a presentation by a student who was professionally dressed and well-prepared. The other group saw a presentation by the *same* student; however, the student was casually dressed and acted underprepared. The professors were then asked to grade a short paragraph written by the student on a topic unrelated to the presentation. The graders who had watched the "good" presentation scored the written response significantly higher even though both groups graded the same written work. This supports the idea of a "halo effect." Teachers tag some students with a positive label and this tag—the halo—creates positive bias during grading.

This is not surprising. As we get to know our students, we inevitably develop associations with each. Some students are naturally gregarious and participate enthusiastically in class. Some students come to every office hour and work hard to improve. Teachers tend to tag those students with a halo. A student who seems less enthusiastic or motivated does not receive a halo. This study suggests that the student without the halo tag may be unfairly penalized during assessment even if their work is at the same level as their peers.

Other research has identified grading bias based on race and ethnicity, gender, and other factors. Instructors must be aware that grading biases are common. Keeping student work anonymous while grading is the simplest strategy to reduce bias. Anonymity can be achieved by having the students write their names on the opposite side of the page from their written answers. Alternatively, you can use a spare piece of paper to cover the names as you grade the written responses. Most classroom management software allows for anonymous grading of electronic student work.

---

These challenges can be addressed through the effective use of an answer key when grading written work. Your key can be as simple as a copy of the assessment that you have filled out yourself. Before you begin grading, fill out a blank copy of the assessment. Take your time and write what you consider to be a complete response to the prompt. For example:

(1) The largest land plants are much larger than the largest land animal. Explain why land plants do not face the same constraints on body size that land animals do. (Answer should be two to three sentences and use class vocabulary throughout. 4 points.)

*Animal body size is constrained by surface area to volume ratio (SA:V).*

*Since most cells in a large plant are dead at functional maturity, they are not constrained by SA:V in the same way.*

Now break down your ideal answer into the components that you will grade. Identify the words or phrases that provide evidence that the student understands the big idea you are assessing. If there are key terms students should use, indicate those on your key.

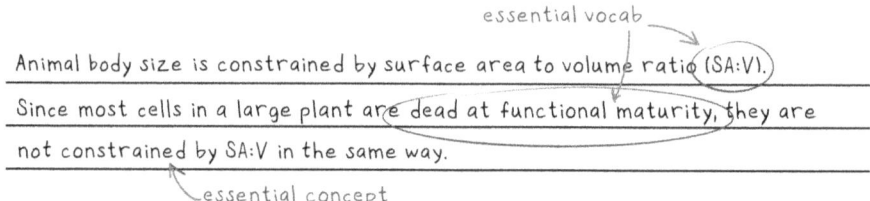

Finally, consider how many points each of those components will be worth and add that information to your key.

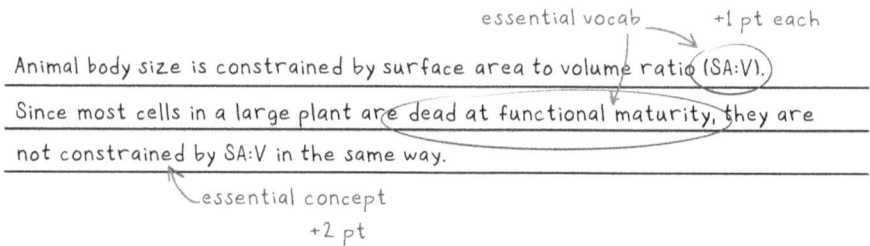

It is a good idea to give your key a test drive before you apply it to a large stack of quizzes. Pull a few quizzes out of the pile and read the student responses through the lens of your key. Do these randomly selected responses all get the same score or are you able to differentiate between responses based on the components you have defined? Do the points assigned to each component make sense? Now is the time to adjust the key rather than realizing halfway through the stack that you need to tweak the point values.

When you are confident that your key is going to be a useful tool, start grading. After looking at several student responses, you may start to notice some trends. For example, you will notice a subset of students misusing an important vocabulary word. Make a note on your key about how you will handle this common error. For example, "If [incorrect term] instead of [correct term] deduct one point." Annotating your key as you grade prevents you from having to try to remember how you scored a particular response earlier so you can quickly and fairly apply the same score to similar responses later.

You might come across some answers that surprise you. Perhaps a student has interpreted the question differently than you anticipated resulting in an alternative answer that could be worth some of the possible points. Add a summary of the response to your key and indicate the number of points you awarded. To continue with the example from earlier, see the note the TA added in the lower right:

> Animal body size is constrained by surface area to volume ratio (SA:V).
> Since most cells in a large plant are dead at functional maturity, they are not constrained by SA:V in the same way.

Annotations on the handwritten key:
- essential vocab, +1 pt each → (SA:V)
- underlined: "dead at functional maturity"
- essential concept, +2 pt → "not constrained by SA:V in the same way"
- "plants don't move – use less energy." +1 pt. maximum

Annotating the key as you go has value beyond efficiency and fairness. If a student comes to your office hours to ask for an explanation of why points were deducted, notes on the key will help you provide an explanation. If you choose to share the key with students as part of a post-quiz review, it will give them insights as they prepare for future assessments—the key shows how an instructor thinks. Your key can also become part of your Teaching Reflection, reminding you of common student mistakes that you might want to address in a future class session.

## Use a Rubric to Grade Essays and Lab Reports

If you are required to grade longer written assignments, planning for efficiency is important to help you stay on top of your responsibilities. Efficiency starts with creating or finding a rubric. A rubric is a guide that spells out your expectations—it acts as a standard unit of measure. The rubric lists the components (ideas, vocabulary, etc.) that a complete response should include. Spending time on a rubric before diving into grading will make grading faster and more fair because every student's response is held to a set standard. Here is an example of a general rubric. Each item you read is given a score from 1 to 4:

| | |
|---|---|
| Student demonstrates complete understanding and uses *discipline-specific language* throughout. | 4 |
| Student demonstrates complete understanding using mostly discipline-specific language. | 3 |
| Student demonstrates *mostly* complete understanding and/or does not use discipline-specific language. | 2 |
| Student demonstrates incomplete understanding. | 1 |

Notice that language use distinguishes the top two scores. The term, "discipline-specific language," refers to vocabulary and ways of speaking common to a particular field. In college-level classes, students should start to use the same vocabulary and expressions used by experts in their chosen area of study. This is most important in classes for majors—a biology class intended for biology majors, for example—and may be somewhat more relaxed in a class for nonmajors where the focus is on exposing students to important concepts without the weight of vocabulary.

The general rubric above can be modified to fit your grading. A single rubric like this can be applied to an entire report or essay. For a report with distinct sections—like the Introduction, Methods, Results, and Conclusions of a lab report—a separate rubric for each section could be developed. Each section receives a score from 1 to 4 and the individual section scores could be added or averaged to yield a final score.

With a rubric to keep your grading efficient and consistent, you are ready to get started on that stack of reports. There is no one-size-fits-all approach but consider the suggestions below and try including some into your grading workflow. Find what works for you.

***Do not overcomment:*** Read the report (or a section of a report) and avoid the urge to start marking it up as you read. In addition to grading, you want to provide feedback to help students improve as writers. You can spend a lot of time writing comments in the margins that overwhelm the students and that may even be ignored. Your feedback will be most valuable if it is concise. If you notice something you might want to come back to and comment on, put a dot in the margin or highlight that text but do not comment for now. When you finish reading, decide what two or three comments will most benefit the student and give those your attention. This is faster for you and more useful for students.

***Go section-by-section:*** Many TAs who grade lab reports find it beneficial to go through the stack one section at a time. They read all the Introductions, scoring each. Then they do the same for the Methods, Results, and Conclusions. Each section needs to be viewed through a different lens by the grader. Grading one section at a time may be less tiring and help improve consistency since you avoid constantly switching your brain from Methods mode to Results mode.

***Do two reports, twice a day:*** This approach is advocated by Professor Daniel Cole. Grade two reports in each sitting and do this twice a day. This way you chip away at the grading pile, rather than doing a marathon grading session. You worry less about the stack of essays and reports knowing that there is a manageable plan in place to work through it. At this pace, you can grade 20 reports in a 5-day week without grading late into the night.

***Set a timer:*** Some TAs put a time limit on each report they grade. With experience you will learn how long it takes you to grade a particular type of assignment. Set a timer on your phone, watch, or computer and do your best to limit all reading and commenting to that interval. If you finish early, you get to take a short break!

## Want to Know More?

Brame CJ. 2013. Writing good multiple choice test questions | Center for Teaching | Vanderbilt University. *Vanderbilt Center for Teaching.* https://cft.vanderbilt.edu/guides-sub-pages/writing-good-multiple-choice-test-questions/

> An online guide to writing multiple-choice questions. Examples illustrating the principles of effective multiple-choice questions are included throughout.

Cole D. 2022. Tackling the stack: tips for grading student essays efficiently and with integrity. *Inside Higher Ed.* https://www.insidehighered.com/advice/2022/04/20/tips-grading-student-essays-efficiently-and-integrity-opinion

*A short article with easy-to-follow suggestions for grading a large pile of essays.*

Crowe A, Dirks C, Wenderoth MP. 2008. Biology in bloom: implementing Bloom's taxonomy to enhance student learning in biology. *CBE Life Sci Educ* 7: 368–381. doi:10.1187/cbe.08-05-0024

*This paper describes ways to use Bloom's taxonomy broadly in teaching to enhance student learning. There are "sentence starters" to help you write—and students recognize—questions at every Bloom's level. This article focuses on biology examples, but the tools can be easily adapted for use in other disciplines.*

APPENDIX

# Guidelines for Teaching

As a TA, you have professional rights and responsibilities that are defined by your institution, and you need to know what these are. You will learn much during your first teaching experience—about yourself as a teacher, about your subject, and about the students in your class. At the same time, you will be expected to learn and follow university and department policies. The goal of this chapter is to help you understand and navigate those policies so you can avoid missteps and be prepared as new situations arise.

Your college or departmental guidelines have probably been made available to you in a document or website. *Ask* if you do not have access to this document and then *read it* when you do have access. Some routine elements of these guidelines are presented here; however, defer to your institutional or departmental policies.

Ultimately, this appendix is all about ***professionalism***, the habits and practices that define an effective teacher. Knowing and being able to communicate content is just one part of teaching. Professionalism—how you interact with students and colleagues and how you represent yourself and your institution—is also part of teaching.

## WORKLOAD AND OFFICE HOURS

Your institution defines the number of hours associated with your teaching position. In addition to in-class time, it is typical for several hours to be required outside of class. The extra required time could be for course coordination, grading, and office hours. Institutions differ in terms of how TAs are compensated for these hours.

As a first-time instructor, you may be required to attend a weekly meeting for the class you are teaching. Coordination meetings may be used to practice effective teaching techniques, give you hands-on experience with lab activities, standardize grading, and provide support to new teachers. Experienced TAs sometimes sit in on coordination meetings even if they are not required to. They find that by sitting in on the coordination meeting and using that time to review and revise their teaching notes, they require little additional time to prepare for teaching. If there are experienced TAs participating in the coordination meetings, take advantage of their experience. They will probably be happy to share advice and teaching materials.

TAs teaching labs or discussion sections might be required to sit in on part or all of an experienced instructor's section. New TAs cite observing an experienced TA as the most important part of their preparation to teach. Make note of how the experienced TA manages time and guides the class. What questions do they ask the class? How do they respond to student questions? These are valuable insights that will help you feel confident as you prepare to teach. TAs teaching as part of a large enrollment class may be required to attend the lecture session. If so, listen for specific language the lead instructor uses or ideas they clearly think are important. Importing ideas and phrases from lecture will reinforce important concepts for students.

Typically, one to two office hours are required for each section that you teach. Office hours are scheduled times when you are available to meet with students—no appointment is needed. Office hours occur in a set location. If you don't have an office or a space to meet with students, your department should provide you with a meeting space when your office hours are in session.

You are probably familiar with the concept of office hours, but students might not be. Early in the term, explain what office hours are and how best to use them. You could say, "Office hours are your time to drop in and get help with the class material. You do not need an appointment. Usually, there are a few students there and we do practice problems, review important concepts, and answer questions."

Your own class and research schedule will determine when you can have office hours. If you are teaching a laboratory class, office hours will be most helpful for students a day or two before your class meets. Students can ask questions about the previous week's lab and any assignments associated with it. They can also ask questions about the upcoming lab to be better prepared. Consider making some of your office hours virtual if allowed by your institution. (Read more about how virtual office hours can promote equality in Chapter 5.)

Being prompt and starting class on time is part of professionalism. Always do your best to be in your classroom ready for an on-time departure. If the hour prior to when you teach is busy—maybe because of your own class schedule—let the students know. Tell them, "I have a class right before this. I'll always do my best to get here as quickly as possible to start our class on time."

## RELATIONSHIPS WITH STUDENTS

Your relationship with students in your class must be clear and appropriate because there are consequences to unprofessional relationships with students that go beyond teaching and learning. As a TA, you may differ in age from the students by just a few years—but students in your class are *not* your peers. Because you have responsibility over their grades, you and the students in your class are not on equal footing in terms of social interactions. (Read more in Chapter 1: Boundaries.)

A romantic or sexual relationship with a student must be absolutely avoided. Your institution has a well-defined policy forbidding "amorous relationships." A typical institutional policy may sound something like this: "Faculty members or other instructional staff shall not initiate, pursue, or be involved in any amorous or sexual relationship with any student they evaluate or supervise by virtue of their teaching, research, or administrative responsibilities."

Let's go further and say that the best policy, guaranteed to keep a TA free of any ethical or legal troubles, is to have *no social contact* with any student in your class. This policy not only protects you, but also avoids the perception of unfairness. A group of students in your class invite you to a party—you should decline. You consider organizing a meet-up with some students in your class outside of office hours—don't do it. Unless *every* student in your class is invited, you should avoid meeting up with students outside of class and office hours. Even if the social interaction is completely innocent, including some students and excluding others—even unintentionally—creates the perception of unfairness. Imagine that before the start of class you are talking with some students about an event you attended together over the weekend. Another student overhearing this would be justified in thinking, "The TA likes them better" or "The TA is their friend."

The excluded students would rightly question your fairness or professionalism.

This advice extends to interactions online. It would not be surprising for students to search for you on social media. If your social media is public, know that students will see your posts. Consider what you are comfortable sharing with this audience. Students may send requests to follow your account. For the same reasons that you should decline an invitation to a social event from a student, decline friend/follow requests from students enrolled in your class.

"WASN'T THAT OUR CHEM TA?"

Of course, if you see students from your class on campus you do not have to run in the opposite direction. Greet them, chat with them, and be friendly. But remember at all times that you are their instructor, and you are in a position of authority. Be friendly toward students while still maintaining professionalism.

## Don't Go Rogue

As a new TA, you are most likely teaching a lab or discussion section that is part of a large-enrollment class with multiple instructors. It is a problem if one TA provides enrichment for their students that students in other sections do not have access to. This could include giving out a study guide, holding review sessions, or posting solution guides to practice problems. The motivation for providing these extras comes from a sincere desire to help the students learn and be successful. However, students talk! Friends in different sections will compare their experiences and feel upset if they aren't given all the resources that others in the same class are provided. You don't want to unintentionally make life harder—or student evaluations worse—for your peers and colleagues. As someone who coordinates a group of TAs teaching sections of the same laboratory class, I have heard the complaints of students who discover that one section has access to helpful study materials that other sections do not have.

"THANKS!"

There is a better way. If you have a great idea that you think will help students learn or will improve the student experience, share it with your coordinator and the other instructors first. Making your innovative teaching tool a shared resource is a win–win scenario. Students benefit from having a resource and the other TAs save time because they do not all need to create the same document individually. Many improvements to labs and lectures have started out as innovative resources brought forward by a TA.

## PRIVACY AND FERPA

Students have a right to privacy in terms of their education. Privacy policies at most universities in the United States are taken from the federal Family Educational Rights and Privacy Act (FERPA). This law states that families and individuals can restrict access to educational records and control the disclosure of personally identifiable information including full names, grades, and pictures. Once a student enters college, the student assumes control of their personal educational information. A parent of a college student is not allowed to see their child's grades unless the student has given permission. It is not unheard of these days for a parent to contact an instructor to ask about the grade status of their child. Such a request prompts questions about parental overreach. However, for the purposes of this chapter, the answer to a parent requesting information about a child is, "No. I can't share information with you without permission from the student." The same goes for a student's friend. It might seem harmless to ask a student to take a graded paper to their absent friend or to say, "Let your friend know they got 92% on the exam." These actions violate FERPA policies. You cannot disclose a student's personal education information without that student's consent.

FERPA guidelines also influence classroom policies. How do you ensure the right to privacy guaranteed by FERPA as you teach your class? Procedures differ from institution to institution, but the following are all compliant with FERPA:

**Do not use full names** if you take attendance ("Lisa" or "Lisa N." is fine). Better yet, if you need to take attendance and you do not know the student names yet, consider walking around the room, checking in with each student individually in the minutes before class starts. This procedure has the added benefit of giving you a one-on-one interaction with each student.

**Use the e-mail or communication tools** in your class's learning management system (LMS, i.e., the class webpage) when you need to e-mail the class or a subset of the class. If you write an e-mail from your personal e-mail account and enter a number of student e-mail addresses, it violates FERPA because all the students included in the e-mail can see the addresses of all the other students. The LMS is designed to let you communicate with a group in a way that does not expose all the names and e-mails. If an LMS is not available, you can use your institutional e-mail as long as you put student e-mails in the blind cc (bcc) field.

**Do not leave graded papers in a public place.** Leaving graded work in a public place for students to pick up may be common practice in some departments, but it violates student privacy. Anyone could look through the pile and see student names and grades. Check before leaving graded papers out in the hallway for students to pick up. Many institutions forbid this. Handing graded papers back in class or office hours is best.

**Do not post lists of student names or grades.** Imagine you are planning an in-class activity with small student groups you set up in advance. To show students what working group they are in, use first names only (include the first letter of the last name in cases where students share the same first name). You can also use the last digits of the students' campus IDs or the first part of their school e-mail address (use "rsmith11" to identify the student if their school e-mail is "rsmith11@college.edu.") None of these ways of identifying students violate FERPA policies.

**Do not post pictures of students without their permission.** This extends to social media and the internet. If you take pictures of students in your class doing a cool chemistry activity, or collecting data in the field, don't share it with the world unless the students have granted you permission to do so.

As you look over your class roster before the start of the term, you may notice that some of the students have opted for *enhanced* FERPA protection. These students have requested a level of privacy around their personal and educational records that goes beyond the basic protections described above. These students may have strong personal reasons for asking for this extra protection. For example, they may want their schedule and location on campus hidden because of a threat of harm from another individual. You will likely not know the full details, but it is good practice to contact students who request enhanced FERPA protections before your class starts—e-mail is fine—to ask some simple questions: "Hello. I see you have requested enhanced FERPA protections. I'll make sure your privacy is respected in the class. I wanted to ask you about some in-class activities. Would you be OK with me calling on you by your first name during discussions? We will be working in groups during class. Can I project your first name on the screen along with the names of the rest of the class when I organize working groups?" Most likely, the student will not have a problem with any of this, but if they do have a need to stay anonymous in class, an e-mail query like this lets you know in advance and helps you plan.

## CHEATING AND PLAGIARISM

Academic dishonesty is hard to deal with and essential to deal with. A case of cheating or plagiarism can be uncomfortable because there is an element of confrontation. You are accusing a student of doing something dishonest. Keep in mind that most students are trying to do their best in the class without looking to gain an unfair advantage over their peers. It is these honest students who you are looking out for. You are acting on their behalf when you deal with a case of suspected academic dishonesty.

You shouldn't be alone when dealing with a case of cheating or plagiarism. If you suspect cheating or plagiarism on a written assignment, show the work to another instructor and get their opinion and support. This isn't required but will help you feel more confident when it is time to discuss your concerns with the student. You can rightly say, "This isn't just my opinion. I showed this to several other instructors, and they agree with my assessment." (When you ask another instructor for their opinion, keep the name of the student anonymous. There are official channels you will use to report academic dishonesty, and no one outside of those channels should know the name of the student involved.)

It is very helpful if the course syllabus explains the procedure for dealing with cases of suspected cheating or plagiarism. If the procedure and consequences are laid out for you and the students, you are not put in the difficult position of making it up as you go. Does a student who plagiarizes a lab report or cheats on an exam get a failing grade in the class, or just on that assignment? At what point does cheating get reported to your institution's office of student rights and responsibilities? If there is not a clear statement about cheating and plagiarism in the syllabus, you can default to a departmental or institutional policy.

If you are teaching one section in a larger course, make sure you know the cheating and plagiarism policy for the course. If there isn't a clear policy, suggest that one be added to the syllabus. If you are responsible for writing your own syllabus, state the consequences of academic dishonesty. Ask around the department for examples or simply refer students to the university's policy. A statement about academic dishonesty might read as follows:

> "Cheating and plagiarism—using other people's work as your own—must be avoided. During exams, always keep your eyes on your own work and do nothing that would give you an unfair advantage over your peers. You will receive no points for any assignment or exam on which cheating or plagiarism has occurred, and you may receive a failing grade in the course at the discretion of the instructor. Cheating will be reported to the Office of Student Rights and Responsibilities and could become part of your academic record."

During a quiz or exam, the potential of a student cheating is reduced when the instructor is attentive. It can be boring to oversee an hour-long exam and it is tempting to try to get other work done or browse on your phone. Your focus should be on the students as they work. Watch them, walk around the room, and keep your eyes open. The students will get the message that you are watching them carefully and the would-be cheater might decide not to risk it. This kind of active exam and quiz proctoring is boring for you. However, being bored for an hour is much better than going through the stress and trouble of following up on a suspected case of cheating. At the start of a quiz or exam, say, "You must keep your eyes on your own work. If I see you looking around, I'll collect your exam and you will get a score of 0." If you notice a student glancing at another's work, repeat those instructions to the *whole* group, "This is your only warning. If I see any student looking at another's work, that student will not be allowed to complete the exam." If the student continues to look around in a suspicious way, follow through and collect that student's exam.

## ACCIDENTS AND INJURIES

This section is primarily for TAs leading laboratory sections. (If you are leading a lecture or discussion section and you and the students are experiencing lots of injuries, you may need to rethink your approach.) If you are leading a lab, you have certain responsibilities regarding student safety. Knowing what the procedures are and whom to contact in case of an injury is information you want to have *before* an injury occurs. Use the pages at the end of the appendix to write down this information. Having this information will ensure that students get the care they need and that you fulfill your responsibilities.

If an accident or injury occurs, the safety of the injured student is the top concern. Err on the side of caution and call for help (911 in the United States) if there is any chance that the injury is serious. If the injury doesn't require an emergency response but does require treatment, you should accompany the student to the health center or ask another student in the class to accompany them. If it is a smaller injury that can be treated with the first aid equipment in the class (and you should know where that is), follow up with the student later by e-mail to check on them.

Your institution has Health and Safety policies and there is very likely to be a designated person at the university or in your department that you need to report injuries to. If an injury has occurred, even a minor one, your first action upon finishing your class is to give a verbal or written report to the designated Health and Safety official. Even if you feel the injury is very minor (a cut finger, a skinned knee on a class hike), let the official know.

## STUDENTS IN CRISIS

Hopefully, the students in your class will stay happy and healthy throughout the term. Given the pressures of academics and the complex social environments of college, it is likely that a few students you work with will experience some crisis and will need help. Before describing scenarios and how best to handle them, remember that *as a TA you are not expected to solve every problem and provide counseling and guidance.* Students in crisis need expert support and there are people on your campus trained to provide that help. That said, if a student shares information with you about personal struggles, responding in a thoughtful way might encourage that student to take the next steps toward a better situation.

Working with students in smaller lab sections or discussion sections allows TAs to notice changes in students—their appearance, behavior, or attendance—that could signal a crisis but that would not be noticed in a large lecture hall. You might notice a normally talkative student suddenly becoming quiet and withdrawn, a student showing up in uncharacteristically disheveled clothing, or a student with a noticeable injury. Such changes could be the outward sign of a crisis.

Another reason that TAs could become aware of a student in crisis earlier than other campus officials is that students could be more willing to confide in a TA. Students tend to see TAs as more approachable, relatable, and understanding than a professor (Kendall and Schussler 2012) and this perception could make a student more willing to reach out to a TA if they are struggling.

If you notice something that concerns you, ask the student how they are doing. For example, if a student has missed more than one class session or has stopped turning in assignments, consider e-mailing the student. "I haven't seen you in class recently and I just wanted to check in and see if you are OK. Let me know if I can support you. My office hours are... ." Depending on the student's response you might put them in touch with campus support services (for mental health, housing, food insecurity, etc.). If you get no response, or if you get a response that raises your level of concern, many campuses have a way to perform a "wellness check" for a student. Which office conducts these wellness checks differs from campus to campus, but it is often the Office of the Dean of Students or the Counseling Center. Requesting a check has the added benefit that now you are not the only person on campus concerned about the student. Others—better trained than you to help a student—are in the loop. Remember, it is better to check and find out that the student is fine than not check and discover later that there was a crisis.

In the case of a very serious crisis involving sexual assault, domestic or partner violence, stalking, or hate crimes, your role becomes very important and slightly more complex. In the United States, colleges that receive federal funding must follow the guidelines of Title IX regarding cases of discrimination and crime on or near campus. Any representative of a Title IX institution, including a TA, is a ***mandatory reporter*** required to disclose information to the appropriate campus authority. This means you cannot promise a student that if they share something with you that you will keep it private. (The designated Title IX official at your campus, who is trained to handle these situations, is able to work confidentially with the student.) Let us examine a student scenario that brings many of the issues described above into play:

A student shows up to class after missing the past two sessions. They seem very tired and quiet. During the session, you ask them to check in with you at the end of class. The conversation might go something like this:

> **TA:** "It's good to see you back in class. How are you doing?"
>
> **Student:** "OK. I know I've missed a few classes and assignments but I'm dealing with a lot, and it's really affected my schoolwork."
>
> **TA:** "I'm sorry you're having a hard time. This thing you are dealing with—are you getting help? Does anyone here at the university know about it?"
>
> **Student:** "No, I'm just dealing with it."

Up until this point, the student has not shared any details with you that you would be required to report. A good approach at this point would be to share a list of the campus offices that can help. (You can use the worksheet at the end of this chapter to organize that information.) The student might not be ready to disclose what's wrong to you, but perhaps they can utilize the list you provide to get help on their own.

> **TA:** "Would you like me to help you get in touch with any of these folks? I know they'd be happy to help you."
>
> **Student:** "I'm just feeling really bad about something that happened to me a few weeks ago. Can I talk to you about it?"

Now it sounds like the student is about to share something significant. Your impulse may be to say anything to help the student feel at ease. However, you need to remind yourself and tell the student that you may need to report what they tell you. Do not promise to keep the conversation private. Do not say "You can tell me anything! I just want to help you." Instead, politely stop the student and tell them about your reporting responsibilities.

> **TA:** "Let me pause you for a second. I really want to listen to you and help you, but you should know that I have an obligation to report what you tell me if there's a chance a crime has been committed. If you choose to tell me what happened, I'll put you in touch with people who can help, and they can keep your information confidential."

By saying this you have disclosed to the student that you have an obligation to report. The student may or may not choose to continue. If they do continue to share their story, listen attentively without interrupting and without judgment. When they are done, say, "I'm sorry that happened to you. Can I walk with you to [campus police, the health center, the Title IX office]." If the student decides not to continue telling their story, let them know, "I understand. If you change your mind, I'll be here and I'm eager to help." You might follow up by e-mailing them later that day to reiterate your offer to help and to share your list of campus resources so they can make contact if they wish to.

## DEPARTMENTAL AND INSTITUTIONAL INFORMATION

There are some important pieces of information that an instructor needs to have. The following pages will remind you of this information and give you a place to record it. This information is specific to your institution or department. Ask for the following information if it has not been provided. Writing the information here will make it accessible if you need it.

## WORKLOAD AND CAMPUS SUPPORT

**Number of required office hours** (per section/per week):

---

**Accident and Injury reporting** (especially for laboratory TAs)

What number do you call in the case of an accident or injury requiring urgent care? (This is probably 911 for U.S. institutions.)

Whom do you call in your department to report the injury?

Who else needs to be notified?

How do you report an injury or accident?

---

**Name, number, and e-mail of the campus office(s) that helps students in crisis.**

Mental health:

Food and housing security:

Initiating a wellness check:

Title IX office (to report a possible crime)

---

**Reporting harassment:** (If you are being harassed by a student or if you witness harassment between members of the campus community.)

## ACADEMIC DISHONESTY

**Penalty for cheating and plagiarism** (For the class you are teaching, how does academic dishonesty affect student grades?)

**Reporting cheating or plagiarism** (Who do you contact when you suspect academic dishonesty? What is the policy for reporting cheating to the University?)

**Contact information for Office of Student Rights and Responsibilities** (help with cheating and plagiarism):

### Want to Know More?

https://studentprivacy.ed.gov

*This site is maintained by the U.S. Department of Education and provides explanations of FERPA rights and responsibilities. There is a helpful section for frequently asked questions.*

Rivenburg R. 2022. *What is Title IX?* University of California, 9 June 2022, https://www.universityofcalifornia.edu/news/what-title-ix. Accessed 13 September 2023.

*A summary of the history and implications of Title IX at colleges and universities.*

# Your Teaching Reflections

Reflecting on your teaching is one of the best ways to improve as a teacher. It is also a great way to save yourself time and energy in the future. The following pages give you space to write a weekly teaching reflection and guidance for what to include. There is one page for each week of a 15-week term. As discussed in Chapter 3, it is best to write a reflection as soon as possible after you finish teaching. Your memory of the details will start to fade as time goes by, so write it down while it is still fresh in your mind. The audience for these reflections is you! Write down anything that will be helpful to you as you prepare to teach this class again in the future. (There is a sample reflection for an introductory laboratory class on page 30 in Chapter 3 to give you a sense for the sorts of details you'll want to write down.)

| |
|---|
| **Date or Week of the Term:** |
| **Name of Lab or Topic of Discussion:** |
| What went well this week? |
| What from this week needs improvement? |
| What technical or management issues arose in this session? How will they be addressed? |
| What questions did students ask? |
| Additional Notes: |

| |
|---|
| **Date or Week of the Term:** |
| **Name of Lab or Topic of Discussion:** |
| What went well this week? |
| What from this week needs improvement? |
| What technical or management issues arose in this session? How will they be addressed? |
| What questions did students ask? |
| Additional Notes: |

| |
|---|
| **Date or Week of the Term:** |
| **Name of Lab or Topic of Discussion:** |
| What went well this week? |
| What from this week needs improvement? |
| What technical or management issues arose in this session? How will they be addressed? |
| What questions did students ask? |
| Additional Notes: |

| |
|---|
| **Date or Week of the Term:** |
| **Name of Lab or Topic of Discussion:** |
| What went well this week? |
| What from this week needs improvement? |
| What technical or management issues arose in this session? How will they be addressed? |
| What questions did students ask? |
| Additional Notes: |

| |
|---|
| **Date or Week of the Term:** |
| **Name of Lab or Topic of Discussion:** |
| What went well this week? |
| What from this week needs improvement? |
| What technical or management issues arose in this session? How will they be addressed? |
| What questions did students ask? |
| Additional Notes: |

| |
|---|
| **Date or Week of the Term:** |
| **Name of Lab or Topic of Discussion:** |
| What went well this week? |
| What from this week needs improvement? |
| What technical or management issues arose in this session? How will they be addressed? |
| What questions did students ask? |
| Additional Notes: |

| |
|---|
| **Date or Week of the Term:** |
| **Name of Lab or Topic of Discussion:** |
| What went well this week? |
| What from this week needs improvement? |
| What technical or management issues arose in this session? How will they be addressed? |
| What questions did students ask? |
| Additional Notes: |

| |
|---|
| **Date or Week of the Term:** |
| **Name of Lab or Topic of Discussion:** |
| What went well this week? |
| What from this week needs improvement? |
| What technical or management issues arose in this session? How will they be addressed? |
| What questions did students ask? |
| Additional Notes: |

| **Date or Week of the Term:** |
|---|
| **Name of Lab or Topic of Discussion:** |
| What went well this week? |
| What from this week needs improvement? |
| What technical or management issues arose in this session? How will they be addressed? |
| What questions did students ask? |
| Additional Notes: |

| |
|---|
| **Date or Week of the Term:** |
| **Name of Lab or Topic of Discussion:** |
| What went well this week? |
| What from this week needs improvement? |
| What technical or management issues arose in this session? How will they be addressed? |
| What questions did students ask? |
| Additional Notes: |

| |
|---|
| **Date or Week of the Term:** |
| **Name of Lab or Topic of Discussion:** |
| What went well this week? |
| What from this week needs improvement? |
| What technical or management issues arose in this session? How will they be addressed? |
| What questions did students ask? |
| Additional Notes: |

| |
|---|
| **Date or Week of the Term:** |
| **Name of Lab or Topic of Discussion:** |
| What went well this week? |
| What from this week needs improvement? |
| What technical or management issues arose in this session? How will they be addressed? |
| What questions did students ask? |
| Additional Notes: |

| |
|---|
| **Date or Week of the Term:** |
| **Name of Lab or Topic of Discussion:** |
| What went well this week? |
| What from this week needs improvement? |
| What technical or management issues arose in this session? How will they be addressed? |
| What questions did students ask? |
| Additional Notes: |

| Date or Week of the Term: |
|---|
| Name of Lab or Topic of Discussion: |
| What went well this week? |
| What from this week needs improvement? |
| What technical or management issues arose in this session? How will they be addressed? |
| What questions did students ask? |
| Additional Notes: |

| |
|---|
| **Date or Week of the Term:** |
| **Name of Lab or Topic of Discussion:** |
| What went well this week? |
| What from this week needs improvement? |
| What technical or management issues arose in this session? How will they be addressed? |
| What questions did students ask? |
| Additional Notes: |

# Bibliography

Addy TM, Dube D, Mitchell KA, SoRelle ME. 2021. *What inclusive instructors do: principles and practices for excellence in college teaching*. Stylus Publishing, London.

Addy TM, Mitchell KA, Dube D. 2021. A tool to advance inclusive teaching efforts: the "Who's in Class?" form. *J Microbiol Biol Educ* **22:** e00183-21.

Allen D, Tanner K. 2005. Infusing active learning into the large-enrollment biology class: seven strategies, from the simple to complex. *Cell Biol Educ* **4:** 262–268.

Anderson LW, Krathwohl DR, Airasian P, Cruikshank K, Mayer R, Pintrich P, Raths J, Wittrock M. 2000. *A taxonomy for learning, teaching, and assessing: a revision of Bloom's taxonomy of educational objectives*. Pearson, New York.

Becker EA, Easlon EJ, Potter SC, Guzman-Alvarez A, Spear JM, Facciotti MT, Igo MM, Singer M, Pagliarulo C. 2018. The effects of practice-based training on graduate teaching assistants' classroom practices. *CBE Life Sci Educ* **16:** ar58.

Bernstein-Yamashiro B, Noam GG. 2013. Establishing and maintaining boundaries in teacher-student relationships. *New Dir Youth Dev* **2013:** 69–84.

Blosser PE. 1991. *How to ask the right questions*. National Science Teachers Association, Arlington, VA.

Blunt S, Hsia K, Rocha A, Shea D. 2018. *Teaching inclusively: quick tips*. Mathematical Association of America. http://digitaleditions.walsworthprintgroup.com/publication/?m=7656&i=475117&view=articleBrowser&article_id=3009234&ver=html5

Bonwell CC, Eison JA. 1991. *Active learning: creating excitement in the classroom*, 1st ed. Washington University Press, Washington, DC.

Boring A. 2017. Gender biases in student evaluations of teaching. *J Public Econ* **145:** 27–41. https://doi.org/10.1016/j.jpubeco.2016.11.006

Brame CJ. 2013. *Writing good multiple choice test questions*. Vanderbilt Center for Teaching, Vanderbilt University. https://cft.vanderbilt.edu/guides-sub-pages/writing-good-multiple-choice-test-questions/

Brown BA. 2019. *Science in the city: culturally relevant STEM education*. Harvard Education Press, Cambridge, MA.

Burton SJ, Sudweeks RR, Merrill PF, Wood B. 1991. *How to prepare better: multiple-choice test items: guidelines for university faculty*. Brigham Young University Testing Services and The Department of Instructional Science, Provo, UT.

Canning EA, LaCosse J, Kroeper KM, Murphy MC. 2020. Feeling like an imposter: the effect of perceived classroom competition on the daily psychological experiences of first-generation college students. *Soc Psychol Pers Sci* **11:** 647–657. https://doi.org/10.1177/1948550619882032

Carriveau RS. 2023. *Connecting the dots: developing student learning outcomes and outcomes-based assessments*, 2nd ed. Routledge, New York.

Carter DJ, Wilson R. 1994. *Minorities in higher education, Annual report*, 13th ed. American Council on Education, Washington, DC.

Chávez K, Mitchell KMW. 2019. Exploring bias in student evaluations: gender, race, and ethnicity. *Pol Sci Pol* 53: 270–274.

Clery Resource Library. Clery Center, https://www.clerycenter.org/resources.

Cole D. 2022. Tackling the stack: tips for grading student essays efficiently and with integrity. *Inside Higher Ed*. https://www.insidehighered.com/advice/2022/04/20/tips-grading-student-essays-efficiently-and-integrity-opinion

Cooper KM, Schinske JN, Tanner KD. 2021. Reconsidering the share of a Think–Pair–Share: emerging limitations, alternatives, and opportunities for research. *CBE Life Sci Educ* 20: fe1.

Crowe A, Dirks C, Wenderoth MP. 2008. Biology in bloom: implementing Bloom's taxonomy to enhance student learning in biology. *CBE Life Sci Educ* 7: 368–381. doi:10.1187/cbe.08-05-0024

Dallimore EJ, Hertenstein JH, Platt MB. 2019. Leveling the playing field: how cold-calling affects class discussion gender equity. *J Educ Learn* 8: 14.

Dewsbury B, Brame CJ. 2019. Inclusive teaching. *CBE Life Sci Educ* 18: 2.

Family Educational Rights and Privacy Act (FERPA). https://studentprivacy.ed.gov

Flaherty C. 2019. Professors' reflections on their experiences with 'ungrading' spark renewed interest in the student-centered assessment practice. *Inside Higher Ed*. https://www.insidehighered.com/news/2019/04/02/professors-reflections-their-experiences-ungrading-spark-renewed-interest-student

Freeman S, Eddy SL, McDonough M, Smith MK, Okoroafor N, Jordt H, Wenderoth MP. 2014. Active learning increases student performance in science, engineering, and mathematics. *Proc Natl Acad Sci* 11: 8410–8415.

Kelly GJ. 2014. Discourse practices in science learning and teaching. In *Handbook of research on science education*, Vol. II, ed. Lederman NG, Abell SK, pp. 321–336. Taylor & Francis Group, New York.

Kendall KD, Schussler EE. 2012. Does instructor type matter? Undergraduate student perception of graduate teaching assistants and professors. *CBE Life Sci Educ* 11: 187–199.

Khazan E, Borden J, Johnson S, Greenhaw L. 2019. Examining gender bias in student evaluations of teaching for graduate teaching assistants. *NACTA J* 64: 422–427. https://www.jstor.org/stable/27157815

Kreitzer RJ, Sweet-Cushman J. 2021. Evaluating student evaluations of teaching: a review of measurement and equity bias in SETs and recommendations for ethical reform. *J Acad Ethics* 20: 73–84.

Lambert LM, Tice SL (eds). 1993. *Preparing graduate students to teach: a guide to programs that improve undergraduate education and develop tomorrow's faculty*. American Association for Higher Education, Grandview, MO.

Lane KA, Meaders CL, Shuman JK, Stetzer MR, Vinson EL, Couch BA, Smith MK, Stains M. 2021. Making a first impression: exploring what instructors do and say on the first day of introductory STEM courses. *CBE Life Sci Educ* 20: ar7. doi:10.1187/cbe.20-05-0098

Lemov D. 2021. *Teach like a champion 3.0: 63 techniques that put students on the path to college*. John Wiley, Hoboken, NJ.

Levin RN. Why asking students their preferred pronoun is not a good idea (opinion). *Inside Higher Ed.* https://www.insidehighered.com/views/2018/09/19/why-asking-students-their-preferred-pronoun-not-good-idea-opinion

Llorens A, Tzovara A, Bellier L, Bhaya-Grossman I, Bidet-Caulet A, Chang WK, Cross ZR, Dominguez-Faus R, Flinker A, Fonken Y, et al. 2021. Gender bias in academia: a lifetime problem that needs solutions. *Neuron* **109**: 2047–2074.

LSE Resources. Evidence-based teaching guides. Inclusive teaching. https://lse.ascb.org/evidence-based-teaching-guides/inclusive-teaching/.

Luckie DB, Aubry JR, Marengo BJ, Rivkin AM, Foos LA, Maleszewski JJ. 2012. Less teaching, more learning: 10-yr study supports increasing student learning through less coverage and more inquiry. *Adv Physiol Educ* **36**: 325–335.

Malouff JM, Emmerton AJ, Schutte NS. 2013. The risk of a halo bias as a reason to keep students anonymous during grading. *Teach Psychol* **40**: 233–237. https://doi.org/10.1177/0098628313487425

Mintzes JJ, Walter EM (eds). 2020. *Active learning in college science: the case for evidence-based practice.* Springer, New York.

Muzaka V. 2009. The niche of Graduate Teaching Assistants (GTAs): perceptions and reflections. *Teach Higher Educ* **14**: 1–12.

NAPE. National Alliance for Partnerships in Equity. http://www.napequity.org.

Orozco D, Lawrence T, Paz-Mondesi Q, Olwell R. 2023. Helping our students understand and fight imposter syndrome. *Faculty Focus.* https://www.facultyfocus.com/articles/effective-classroom-management/helping-our-students-understand-and-fight-imposter-syndrome/.

Orr RB, Csikari MM, Freeman S, Rodriguez MC. 2022. Writing and using learning objectives. *CBE Life Sci Educ* **21**: fe3.

Palmer C. 2021. How to overcome impostor phenomenon. *Monit Psychol* **52**: 44.

Pronouns.org. Resources on personal pronouns. How do I share my personal pronouns? https://pronouns.org/sharing.

Rivenburg R. 2022. *What is Title IX?* University of California. https://www.universityofcalifornia.edu/news/what-title-ix

Rockquemore KA. 2015. How to listen less. *Inside Higher Ed.* https://www.insidehighered.com/advice/2015/11/04/setting-boundaries-when-it-comes-students-emotional-disclosures-essay

Samuels DR. 2014. *The culturally inclusive educator: preparing for a multicultural world.* Teachers College Press, New York.

Schussler E, Torres LE, Rybczynski S, Gerald GW, Monroe E, Sarkar P, Shahi D, Osman MA. 2008. Transforming the teaching of science graduate students through reflection. *J Coll Sci Teach* **38**: 32–36.

Schwartz HL. 2020. Role clarity: how faculty can map their own boundaries. *NEA Higher Education Advocate*, Vol. 8. https://www.nea.org/sites/default/files/2020-08/How%20faculty%20can%20map%20their%20own%20boundaries.pdf

Tanner KD. 2009. Talking to learn: why biology students should be talking in classrooms and how to make it happen. *CBE Life Sci Educ* **8**: 89–94.

Tanner KD. 2013. Structure matters: twenty-one teaching strategies to promote student engagement and cultivate classroom equity. *CBE Life Sci Educ* **12**: 322–331.

The American Association for the Advancement of Science. 2015. *Vision and change in undergraduate biology education: chronicling change, inspiring the future.* AAAS, Washington, DC. www.visionandchange.org.

Townsend BK, Wilson KB. 2009. The academic and social integration of persisting community college transfer students. *J Coll Stud Ret* **10**: 405–423.

University of California, Riverside School of Medicine. *Writing effective test questions*. School of Medicine Faculty Development. https://facdev.ucr.edu/writing-effective-test-questions

U.S. Bureau of Labor Statistics. 2023. 25-9044. Occupational employment and wages, May 2022. Teaching Assistants, Postsecondary, Bureau of Labor Statistics, Washington, DC. https://www.bls.gov/oes/current/oes259044.htm#(1)

Wyrick A. 2022. *How to define boundaries with your students—and stick to them*. Harvard Business Publishing, Cambridge, MA. https://hbsp.harvard.edu/inspiring-minds/how-to-define-boundaries-with-your-students-and-stick-to-them

# Index

**A**
Academic dishonesty, 108–109, 114
Accidents, 109–110, 113
Active learning, 59–74
    asking good questions, 64–67
    classroom design for, 70–71
    forms of in-class participation, 73
    in a lab: negative controls, 63
    in a lecture: ideal gas law, 61–62
    passive learning compared, 60
Adding students, 15
Address, preferred form of, 8, 21
Amorous relationships, 104–105
Apology, 10
Assessments, 85–101
    Bloom's taxonomy, 86–90, 92–96
    challenges of, 85–86
    cheating and plagiarism during, 108–109
    writing, 89–96
    *See also* Grading
Attendance, taking, 20, 107

**B**
Bias
    boundaries, 11–12
    in grading, 85–86, 97
    halo effect, 97
Bloom's taxonomy, 86–90, 92–96
Board game analogy, 5–8
Boundaries, 5–12
    bias and, 11–12
    board game analogy, 5–8
    conversations, advice about, 10–11
    defining, 8–9
    in-the-moment decision about, 9–10
    maintaining, 9–11
    "Never OK!" behaviors, 7–8
    student challenges to, 11–12

**C**
Campus resources, 49
Change, advocating for, 50
Cheating, 108–109, 114
Chemistry quiz, crafting, 94–96
Closed questions, 65
Cold calling, 72
Collaboration, encouraging, 52
Competition, 49, 52
Content, 31–42
    case study: fruit lab, 36–37
    guiding student learning, 32–33
    introducing in first class, 21
    sharing your learning process, 34–36, 37
    steps in preparation, 33–34
    system for preparing new, 37–38
    week-to-week, preparing and planning, 41–42
Content preparation guide, 39–40
Content preparation steps
    building confidence in minimal material, 34, 36
    building knowledge, 34, 37
    identify the minimum, 33–34, 36
    staying alert for insights, 34, 36–37
Coordination meetings, 103
Crashers, 15
Crisis, students in, 110–112
Current events, acknowledging, 48

**D**
Departmental and institutional information, 112–114
Discrimination, 111

Distractors, 90–91, 94–95
Diversity, 45, 47, 50, 56
Diversity statement, 50

## E
E-mail
    privacy, 107
    welcome, 23–24, 49–50
Equity, 43, 45, 50
Error, normalizing, 68–70
Essays, grading, 99–100
Exclusion, 43–45, 54–55, 105

## F
Family Educational Rights and Polices Act (FERPA), 20, 106–108
Feedback, on public speaking, 20
Fill-in-the-blank questions, 91
First class, 13–24
    being welcoming, 20–21
    crashers, 15
    first-day concerns, 23–24
    giving students chance to talk, 22–23
    inclusive language, 21
    speaking in the classroom, 19–20
    writing in class, 19
Flexibility, 48
Formal assessment, 85
Free-response questions, 92–94

## G
Gender bias, 11–12
Grading, 85–101
    anonymous, 97
    bias, 85–86, 97
    earned grades, 86
    essays and lab reports, 99–100
    fair, efficient, 96–100
    posting grades, 107
    privacy issues, 107
    suggestions for workflow, 100
    writing comments for feedback, 100
    written responses, 96–99
Guide for student learning, being, 32–33
Guidelines for teaching, 103–114
    accidents and injuries, 109–110, 113
    cheating and plagiarism, 108–109, 114
    privacy and FERPA, 106–108
    providing extra resources, 105–106
    relationships with students, 104–106
    students in crisis, 110–112
    workload and office hours, 103–104

## H
Halo effect, 97
Hand signals, 68
How-Do-You-Know questions, 66

## I
Imposter feelings
    raising awareness of, 51–52
    student, 21, 46, 48, 51–52
    teaching assistant, 24, 32, 52
Inclusive language, 21
Inclusivity, 43–58
    being an inclusive teacher, 44–46
    exclusionary language from students, 54–55
    imposter feelings, 46, 48, 51–52
    mindset, 46
    preterm survey, 50, 56–57
    pronouns, 47, 55
    taking action, 47–50
    words, influence of, 52–54
Inequity, 43–44, 53–54
Injury, 109–110, 113
Instructor guide, 14, 16
Intentional actions, 47–50
Intentionality, 45
Introducing yourself, 21

## K
Key, 96–99

## L
Lab reports, grading, 99–100
Large-enrollment classes, inclusivity in, 46
Learning management system (LMS), 107
Learning objectives, 89
Learning process, sharing your, 34–36, 37

## M
Managerial questions, 64–65, 67
Mandatory reporter, 111
Microinequities, 53–54
Microinflammations, 53–54

Mindset
   inclusive, 46
   positive, 1–2
Multiple-choice questions, 90–92, 93

**N**
Names of students
   learning, 47
   privacy, 107–108
Negative phrasing, 91
Normalizing error, 68–70
Notetaking apps, 19

**O**
Office hours
   evening, 49
   guidelines, 104
One-minute notes, 73
Open questions, 65

**P**
Parents, 106
Participation, 59–74
   asking good questions, 64–67
   cold calling, 72
   encouraging in other students, 71
   first-day concerns, 23–24
   hand signals, 68
   inclusivity, 48
   mobility in the classroom, 70–71
   normalizing error, 68–70
   reducing barriers to, 48
   silence as tool for, 66–67
   Think–Pair–Share (TPS), 75–83
Passive learning, 59–63
   active learning compared, 60
   classroom scenarios switching to active learning, 61–64
Personal education information, 106
Personal Teaching Document, 14, 16–17, 36
Physical space, familiarity with, 18
Pictures of students, 107
Plagiarism, 108–109, 114
Positive tone, setting, 21
Practice walk-through, 18
Preparation
   first class, 13–20
   steps in, 33–34

Preterm survey, 50, 56–57
Privacy, 106–108
Proctoring, 109
Professionalism, 103
Pronouns, 47, 55
Public speaking, 19–20

**Q**
Questions
   closed, 65
   cold calling, 72
   hand signal answers, 68
   high-stakes and lower-stakes, 65–66
   How-Do-You-Know, 66
   managerial, 64–65, 67
   normalizing error, 68–70
   open, 65
   rhetorical, 64
   during Think–Pair–Share (TPS), 77–78
   wait time, 66–67
Questions, assessment
   Bloom's taxonomy, 86–90, 92–96
   choosing best answer, 91, 93–94
   fill-in-the-blank, 91
   free-response or short-answer, 92–94
   multiple-choice, 90–92, 93
Questions, student
   addressing unanswered, 24
   first-day, 22–23
   valuing, 22
Quizzes. *See* Assessments

**R**
Reflection, 25–30, 115–130
   information to include, 27
   sample, 30
   template, 29, 116–130
   using, 28–29
   writing, 26–27
Relationships with students, 104–106
Respect, from students, 24
Rhetorical questions, 64
Rights and responsibilities, 103
Role model, 43
Rubric, 99–100

**S**
Safety, student, 109–110, 113
SafeZone training, 50

Self-criticism, 25
Sharing your experience, 48
Short-answer questions, 92–94
Silence, as tool for getting participation, 66–67
Social connections, between students, 49
Social contact with students, 105
Social media, 105, 107
Speaking in the classroom, 19–20
Stem, 90–92
Students in crisis, 110–112
Support, 14
Support services, campus, 110–113

**T**
Talking, giving students a chance for, 22–23
Teacher-centered instruction, 45, 59–61, 63–64
Teaching Reflection, 62, 99, 115–130
Test bank, 88–89
Think–Pair–Share (TPS), 23, 48, 75–83
    to address too-long silence, 67
    group size, 77
    overview, 75–76
    Pair phase, 77–78, 82–83
    planner, 79–81
    Share phase, 78–79, 82
    Think phase, 76–77
    variations of, 82–83
Title IX, 6, 8, 111

**U**
Ungrading, 85

**W**
Warm calling, 73
Welcoming students, 20–21, 47
Wellness check, 110
"Who's in Class" form, 56
Workload, 103–104
Writing assessments, 89–96
    chemistry quiz example, 94–96
    free-response or short-answer items, 92–94
    multiple-choice items, 90–92
Writing on the board, 19

# About the Author

Ed Himelblau (he/him) is a biology professor at California Polytechnic State University. He is also a cartoonist for *The New Yorker* and other publications. (See Ed's cartoons at himelblau.com and @himelblog.) Ed lives in San Luis Obispo, California with his wife and daughter. He received the Cal Poly Distinguished Teacher Award in 2018. Since 2015, Ed has taught a class for incoming graduate students who are teaching for the first time.